The Biliary System

Colloquium Series on Integrated Systems Physiology: From Molecule to Function to Disease

Physiology is a scientific discipline devoted to understanding the functions of the body. It addresses function at multiple levels, including molecular, cellular, organ, and system. An appreciation of the processes that occur at each level is necessary to understand function in health and the dysfunction associated with disease. Homeostasis and integration are fundamental principles of physiology that account for the relative constancy of organ processes and bodily function even in the face of substantial environmental changes. This constancy results from integrative, cooperative interactions of chemical and electrical signaling processes within and between cells, organs, and systems. This eBook series on the broad field of physiology covers the major organ systems from an integrative perspective that addresses the molecular and cellular processes that contribute to homeostasis. Material on pathophysiology is also included throughout the eBooks. The state-of the-art treatises were produced by leading experts in the field of physiology. Each eBook includes stand-alone information and is intended to be of value to students, scientists, and clinicians in the biomedical sciences. Since physiological concepts are an ever-changing work-in-progress, each contributor will have the opportunity to make periodic updates of the covered material.

Published titles

(for future titles please see the Web site, www.morganclaypool.com/page/lifesci)

The Biliary System
David Q.-H. Wang, Brent A. Neuschwander-Tetri, and Piero Portincasa
www.morganclaypool.com

ISBN: 9781615043606 paperback

ISBN: 9781615043613 ebook

DOI: 10.4199/C00051ED1V01Y201202ISP033

A Publication in the

COLLOQUIUM SERIES ON INTEGRATED SYSTEMS PHYSIOLOGY: FROM MOLECULE TO FUNCTION TO DISEASE

Lecture #33

Series Editors: D. Neil Granger, LSU Health Sciences Center, and Joey P. Granger, University of Mississippi Medical Center

Series ISSN

ISSN 2154-560X print

ISSN 2154-5626 electronic

The Biliary System

David Q.-H. Wang, Brent A. Neuschwander-Tetri, and Piero Portincasa
Saint Louis University School of Medicine and University of Bari Medical School

INTEGRATED SYSTEMS PHYSIOLOGY: FROM MOLECULE TO FUNCTION TO DISEASE #33

ABSTRACT

The liver is a vital organ involved in numerous metabolic processes such as cholesterol and bile acid metabolism, biliary lipid secretion, and bile formation. Cholesterol balance across the liver has a crucial effect on influencing plasma total and LDL cholesterol levels and biliary cholesterol concentrations. Cholesterol and bile acid biosyntheses are primarily modulated by negative feedback regulation mechanisms through the sterol regulatory element-binding protein isoform 2 (SREBP-2) and the farnesoid X receptor (FXR) pathways, respectively. The hepatic conversion of cholesterol to bile acids can balance the fecal excretion of bile acids, which is an important route for the removal of cholesterol from the body. Bile formation begins in the bile canaliculi, and maintenance of the enterohepatic circulation of bile acids results in a continuous secretion of bile. Hepatic secretion of biliary lipids is determined mainly by a group of ATP-binding cassette (ABC) transporters that are regulated by various nuclear receptors. Bile acids promote bile flow by their osmotic effects. Also, they are essential for the intestinal absorption of cholesterol, fatty acids, and fat-soluble vitamins and play an important role in aiding the digestion of dietary fat. Bile acids function as signaling molecules and anti-inflammatory agents to regulate lipid, glucose, and energy metabolism by rapidly activating nuclear receptors and cell signaling pathways. This eBook summarizes the progress in understanding the molecular mechanism of cholesterol and bile acid metabolism and the physical-chemistry of biliary lipids, with emphasis on biliary lipid metabolism that is regulated by nuclear receptors in the hepatobiliary system.

KEYWORDS

bile, cholesterol synthesis, bile acid metabolism, biliary secretion, enterohepatic circulation, hepatic lipid transporter

Contents

CHAPTER 1
Introduction

The liver plays a central role in the regulation of cholesterol and bile acid metabolism and is involved in biliary lipid secretion and bile formation. Cholesterol is synthesized from acetyl CoA, and 3-hydroxy-3-methylglutaryl CoA (HMG-CoA) reductase is the rate-limiting enzyme for cholesterol biosynthesis in the body. The liver provides a major source of the cholesterol molecules to the body. Cholesterol biosynthesis is primarily modulated by a negative feedback regulation mechanism through the sterol regulatory element-binding protein isoform 2 (SREBP-2) pathway in the liver. The molecular regulation of SREBPs occurs at two levels—transcriptional and post-transcriptional—in the nucleus of hepatocyte. Cholesterol is secreted into bile either in a "free" (unesterified) form or in its end-product form after it is converted to bile acids by the hepatocyte. Hepatic secretion of biliary cholesterol into the bile is an important pathway for the elimination of cholesterol from the body. Furthermore, cholesterol balance across the liver has a crucial effect on influencing plasma total and LDL cholesterol levels and biliary cholesterol concentrations. Because HMG-CoA reductase catalyzes an irreversible step on the synthesis of mevalonate, it plays an important role in cholesterol biosynthesis in the liver. Also, it is a crucial target for the treatment of hypercholesterolemia or dyslipidemia in patients, as found by basic experiments and confirmed by clinical studies with administration of HMG-CoA reductase inhibitors statins.

The bile acid biosynthesis in the liver involves two major pathways: the "classic" neutral and the "alternative" acidic pathways. In the classic neutral pathway, cholesterol 7α-hydroxylase (CYP7A1) is an important enzyme that is expressed only in the hepatocyte and catalyzes the rate-limiting step in the catabolism of cholesterol to bile acids by converting cholesterol directly into 7α-hydroxycholesterol. In the alternative acidic pathway, cholesterol is transported to mitochondria where it must first be converted into oxysterols by a group of sterol hydroxylases. Bile acid synthesis involves at least 17 enzymes that make 25 possible intermediates. Furthermore, CYP7A1 plays a crucial role in the regulation of bile acid biosynthesis in the liver, which is modulated by a negative feedback regulation system through a group of nuclear receptors. The mechanism underlying farnesoid X receptor (FXR) signaling in the inhibition of *CYP7A1* transcription and bile acid synthesis has been investigated extensively. These studies have revealed that there are two FXR-dependent mechanisms for bile acid inhibition of *CYP7A1* gene transcription. In the liver, FXR can inhibit

CYP7A1 via the small heterodimer partner (SHP) pathway. In the intestine, FXR activated by bile acids stimulates the release of fibroblast growth factor 19 (FGF19). FGF19 circulates to the liver and triggers hepatic FGF receptor 4 (FGFR4) signaling to inhibit CYP7A1. In addition, bile acid inhibition of CYP7A1 is determined, in part, by several FXR-independent mechanisms.

The enterohepatic circulation of bile acids is an important physiological route for recycling of bile acids from the liver to the small intestine and back to the liver. Then, bile acids are re-secreted into bile. The recycling of bile acids plays a vital role in the negative feedback regulation of bile acid biosynthesis in the liver. Furthermore, bile acids not only promote bile flow and intestinal absorption of fat, cholesterol, fat-soluble vitamins, and other nutrients, but also work as signaling molecules and inflammatory agents to regulate lipid, glucose, and energy metabolism by rapidly activating nuclear receptors and cell signaling pathways. Of note, the enterohepatic circulation of bile acids is highly efficient, allowing less than 5% of the secreted bile acids to be lost in the feces. Because biliary bile acids are secreted in such large amounts, approximately 0.4 g of bile acids is lost in the feces per day. Thus, the conversion of cholesterol to bile acids, i.e., bile acid synthesis in the liver, can balance the fecal excretion of bile acids, which is also an important route for the elimination of cholesterol from the body.

The bile is produced by the liver and bile formation begins in the bile canaliculi of hepatocytes. On the canalicular membrane of the hepatocyte, an ATP-binding cassette (ABC) transporter ABCB11, a bile acid export pump, promotes the secretion of bile acids into bile. Subsequently, biliary bile acids stimulate two other ATP-dependent transporters—ABCB4 and a heterodimer of ABCG5 and ABCG8—to induce the hepatic secretion of phospholipid and cholesterol into bile, respectively. Bile acids are highly soluble in water and are biological detergents. In contrast, cholesterol and phospholipid are virtually insoluble in water. When a critical micellar concentration (CMC) is exceeded, bile acids can self-assemble into simple micelles in bile. Also, they bind with cholesterol and phospholipid to form mixed micelles. These micelles help solubilize cholesterol in bile. Furthermore, the phospholipid molecules can aggregate to form unilamellar vesicles. These vesicles greatly promote the solubilization of cholesterol in bile and help transport the cholesterol molecule from the liver to the small intestine. Although vesicles are quite static structures, the equilibrium between vesicles and micelles is influenced by several factors such as the total lipid concentration and the relative ratio of cholesterol, phospholipids, and bile acids in bile. These physical structures of biliary lipid carriers have an important effect on bile formation, as well as the digestion and absorption of dietary fat, cholesterol, lipid-soluble vitamins, and some drugs.

This eBook will carefully review the hepatobiliary aspects of cholesterol and bile acid metabolism and physical-chemistry of biliary lipids, as well as summarize current knowledge of molecular regulation of cholesterol and bile acid biosyntheses in the liver, with emphasis on biliary lipid metabolism that is regulated by nuclear receptors in the hepatobiliary system.

• • • •

CHAPTER 2

Anatomy of the Liver, Biliary Tract, and Gallbladder

THE LIVER

Gross and Surface Anatomy

The liver is the largest solid organ in the body. It is wedge-shaped with its base against the right abdominal wall and its tip pointing to the spleen. The liver has two major surfaces: the diaphragmatic and visceral surfaces. The diaphragmatic surface faces anteriorly and superiorly, whereas the visceral surface faces posteroinferiorly. Although most of the liver is covered with a layer of visceral peritoneum, the superior part, called the bare area, is fused to the diaphragm and therefore lacks peritoneum (Figure 2.1). The liver extends from the fifth intercostal space in the right midclavicular line down to the right costal margin, with its size being 12 to 15 cm coronally and 15 to 20 cm transversely. It lies below the diaphragm and occupies most of the right hypochondrial and part of the epigastric regions of the abdomino-pelvic cavity. The liver weight ranges from 1.3 to 1.7 kg (between 1.8% and 3.1% of body weight) in most adults, depending on gender and body size. The liver weight in fetuses and children is relatively greater, being 5.6% at 5 months gestational age, 4% to 5% at birth, and 3% at 1 year of age.

The liver is covered by the fibrous "Glisson's" capsule and has a continuous sponge-like parenchymal mass penetrated by tunnels that contain the interdigitating networks of afferent and efferent vessels [1–5]. The primary afferent blood vessels of the liver are the portal vein and hepatic artery. The branches of these vessels within the liver parenchyma are contained together in connective tissue which is contiguous with the mesenchymal components of the liver's mesothelium-covered surface capsule. The finest branches of the efferent vessels become major components of the portal tracts. Other components of the portal tracts include nonparenchymal cells and cell types that make important contribution to liver function, including bile metabolism (cholangiocytes in bile ducts), vascular regulation (sympathetic nerves), pain perception (parasympathetic nerves), immune function (immunocytes), and lymph formation (lymph vessels). The collagenous stroma surrounding the efferent vessel, from the central veins to the hepatic vein, is less robust and contains fewer adventitial cells.

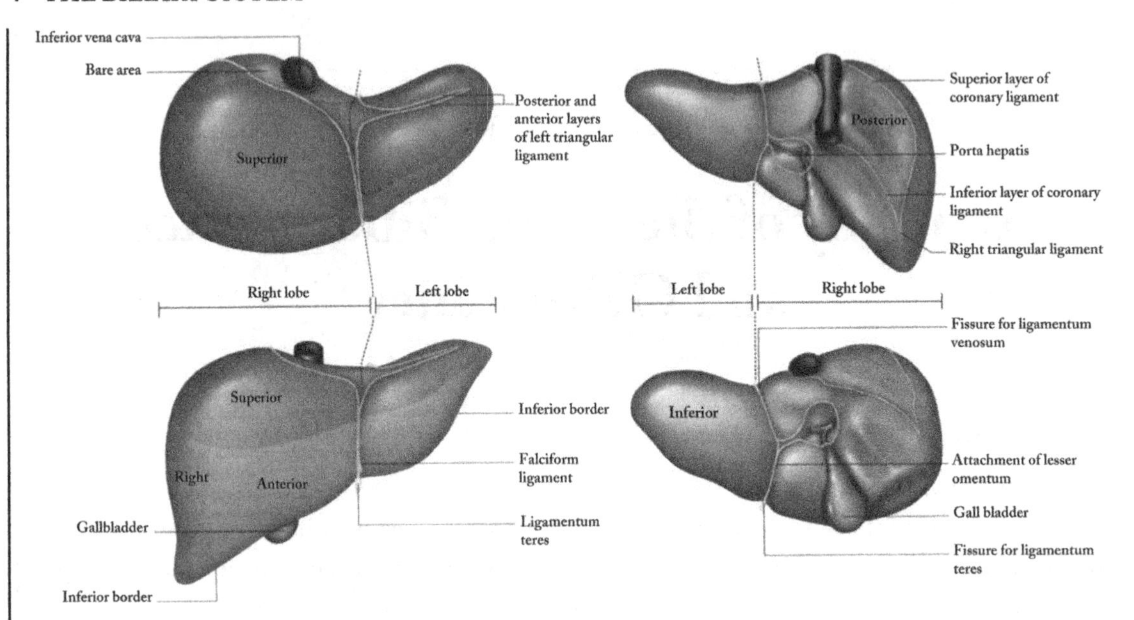

FIGURE 2.1: Gross anatomy of the liver. Top left, superior view; top right, posterior view; bottom left, anterior view; bottom right, inferior view. Used with permission from *Gray's Anatomy: The Anatomical Basis of Clinical Practice.* Editor-in-Chief: Standring S. 39th edition. Elsevier Churchill Livingstone, London, 2005. p. 1214.

At the porta hepatis, defined as the region in which the portal vein and hepatic artery enter the liver and the hepatic bile duct exits the liver, the connective tissue of the capsule is continuous with the fibrous sheath investing the portal vessels and bile ducts and following them to their smallest ramifications [1–5]. At the superior surface of the liver, the capsular peritoneum reflects onto the diaphragm and continues as the parietal peritoneum. The reflections form the coronary ligaments, the right and left triangular ligaments, and the falciform ligament. These ligaments hold the liver in its place and allow the passage of lymphatics, small vessels, and nerves. The falciform ligament, a fold of the parietal peritoneum, extends from the undersurface of the diaphragm between the principal lobes of the liver. The round ligament (ligamentum teres), a fibrous cord containing a remnant of the umbilical vein of the fetus, forms the free border of the falciform ligament. The right and left coronary ligaments are narrow reflections of the parietal peritoneum that suspend the liver from the diaphragm. In addition, there is a large bare area where the liver contacts the diaphragm and retroperitoneum. The vena cava, being retroperitoneal, lies within the bare area and attached to the liver by a ligament or bridge of the liver parenchyma between the caudate and right lobes.

The hepatoduodenal ligament connects the liver to the superior part of the duodenum. It is part of the lesser omentum and sheathes the hepatic artery, the portal vein, bile ducts, nerves, and lymph vessels, all being present within the porta hepatis [6]. In the ligament, the common bile duct lies to the right, the hepatic artery to the left, and the portal vein behind them. Numerous variations in the topography of the hepatic artery are common.

Structural Concepts of Liver Lobes and Segmentation

The human liver is divided into two principal lobes—a large right lobe and a smaller left lobe (Figure 2.2). In general, the falciform ligament on the anterior diaphragmatic surface, and the lesser omentum and umbilical fissure on the posterior visceral surface, divide the liver into the conventional right and left lobes [1–5]. However, this division of the liver does not correspond to the division based on branch points in the vascular supply. The liver can be divided on a different plane into right and left livers (or hemilivers), each with its own blood supply and bile duct drainage. The right

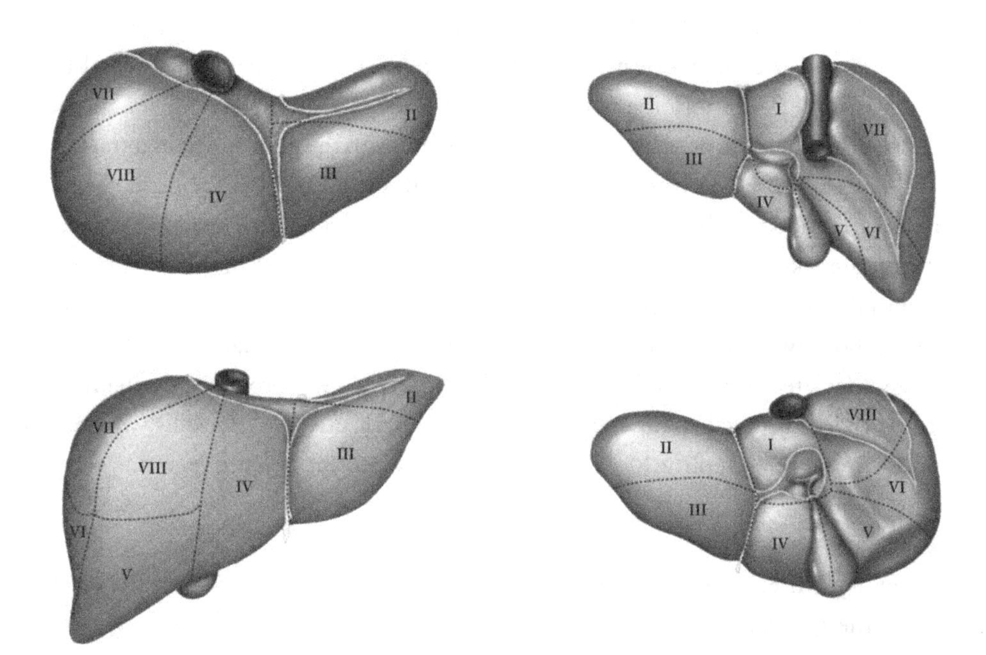

FIGURE 2.2: Segmental anatomy of the liver based on the Couinaud terminology. Eight segments are identified. Top left, superior view; top right, posterior view; bottom left, anterior view; bottom right, inferior view. The segments are sometimes referred to by name—I, caudate (sometimes subdivided into left and right parts); II, lateral inferior; III, lateral inferior; IV, medial (sometimes subdivided into superior and inferior parts); V, anterior inferior; VI, posterior inferior; VII, posterior superior, VIII, anterior superior. Used with permission from *Gray's Anatomy: The Anatomical Basis of Clinical Practice*. Editor-in-Chief: Standring S. 39th edition. Elsevier Churchill Livingstone, London, 2005. p. 1217.

hemiliver makes up 50% to 70% of the liver mass [1–5]. Two other lobes, the smaller quadrate and caudate lobes, are visible on the visceral surface just to the right of the fissure. Although the right lobe is considered to include an inferior quadrate lobe and a posterior caudate lobe, on the basis of internal morphology (primarily the distribution of blood vessels and nerves), the quadrate and caudate lobes more appropriately belong to the left lobe. Germane to surgical resection, the liver can be further divided into a total of eight segments on the basis of the vascular or bile duct distribution. This segmental nomenclature devised by Couinaud (Figure 2.2) has received the widest acceptance because of its relevance to surgical resection of focal lesions of the liver [7]. This classification is based on the divisions of the portal veins. However, the branching of the portal veins to the left lobe is irregular because of the entry of the umbilical vein, making it desirable to adopt a nomenclature based on the divisions of the arteries or ducts [8].

Large Vessels of the Liver

An important area near the center of the visceral surface is the porta hepatis where most of the major vessels and nerves enter and leave the liver (Figure 2.3). The liver receives blood from two afferent vessels (Figure 2.4). From the portal vein, it obtains nutrient-rich blood containing newly absorbed nutrients, drugs, and possibly microbes and toxins from the gastrointestinal tract. Also, the liver receives oxygen-rich blood from the hepatic artery. Branches of both the portal vein and the hepatic artery carry blood to the liver sinusoids, where many of the nutrients, oxygen, and certain toxic substances are taken up by the hepatocytes. About two thirds of the blood flowing to the liver is from the portal vein while the hepatic artery contributes the other third. In contrast, the hepatic artery supplies about two thirds of the oxygen to the liver while the portal vein delivers the other third after been depleted of much of its oxygen. Metabolic products produced by the hepatocytes and nutrients needed by extrahepatic organs are secreted back into the sinusoids, which then coalesce into the central veins and the blood eventually drains into the hepatic vein.

Portal Veins

The portal vein is an afferent nutrient vessel of the liver, which carries blood from the entire capillary system of the gastrointestinal tract, spleen, pancreas, and gallbladder. The portal vein supplies blood to the parenchymal mass through its terminal branches (Figure 2.4). The portal vein is formed behind the neck of the pancreas by the confluence of the splenic vein and the superior mesenteric vein. It also receives the superior pancreaticoduodenal vein, the left gastric (coronary) vein, and the cystic vein.

The splenic vein originates from five to six branches that return the blood from the spleen and merge to form a single vessel. The superior mesenteric vein carries blood from the small intestine, ascending colon, and transverse colon. The inferior mesenteric vein returns blood from the area drained by the superior and the inferior left colic and the superior rectal veins.

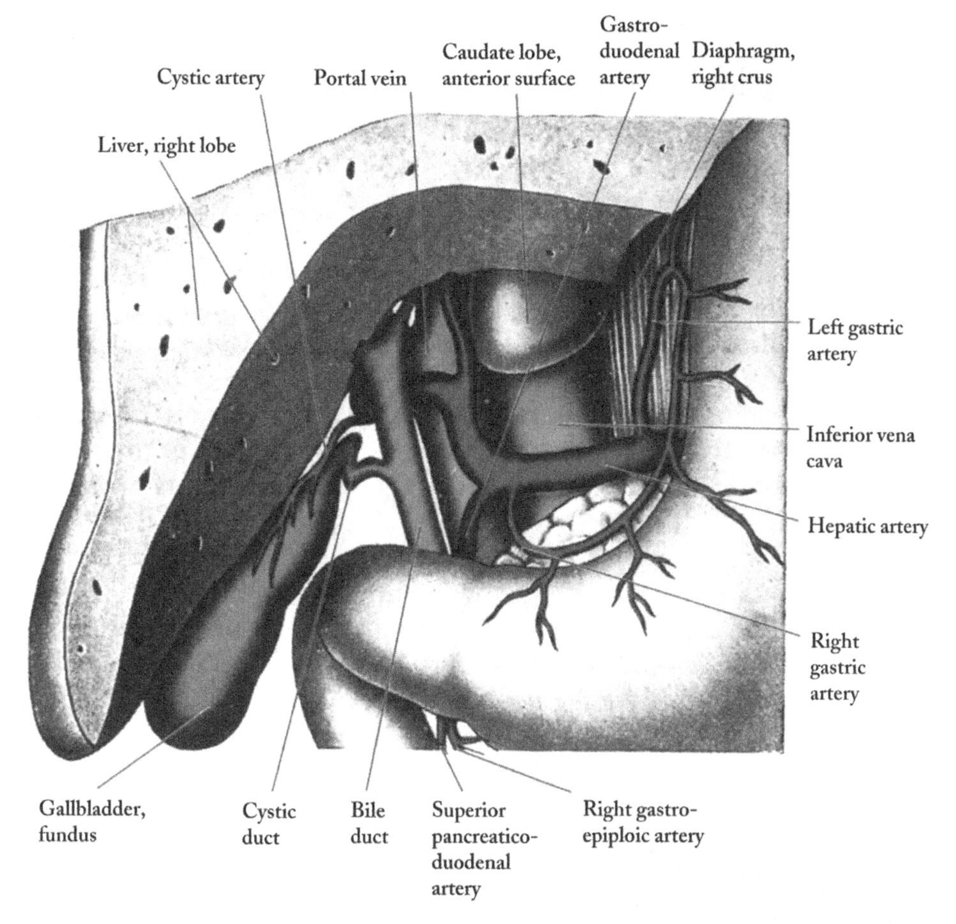

FIGURE 2.3: Anatomy of the porta hepatis, showing the relationship among the common hepatic duct, the cystic duct, the common bile duct, the hepatic artery, and the hepatic portal vein, as well as the Calot's triangle. Used with permission from *Gray's Anatomy: The Anatomical Basis of Clinical Practice*. Editor-in-Chief: Standring S. 39th edition. Elsevier Churchill Livingstone, London, 2005. p. 1218.

The portal trunk runs in the hepatoduodenal ligament in a plane dorsal to the bile duct and the hepatic artery and splits into two lobar veins before entering the portal fissure (Figure 2.4). The right lobar vein, short and thick, receives the cystic vein. The left lobar vein, longer and smaller, is joined by the umbilical vein and the paraumbilical veins.

Hepatic Arteries

The common hepatic artery is the second major branch of the celiac axis. It splits into the right and the left hepatic arteries to supply the corresponding hemilivers (Figure 2.4). The right and the left

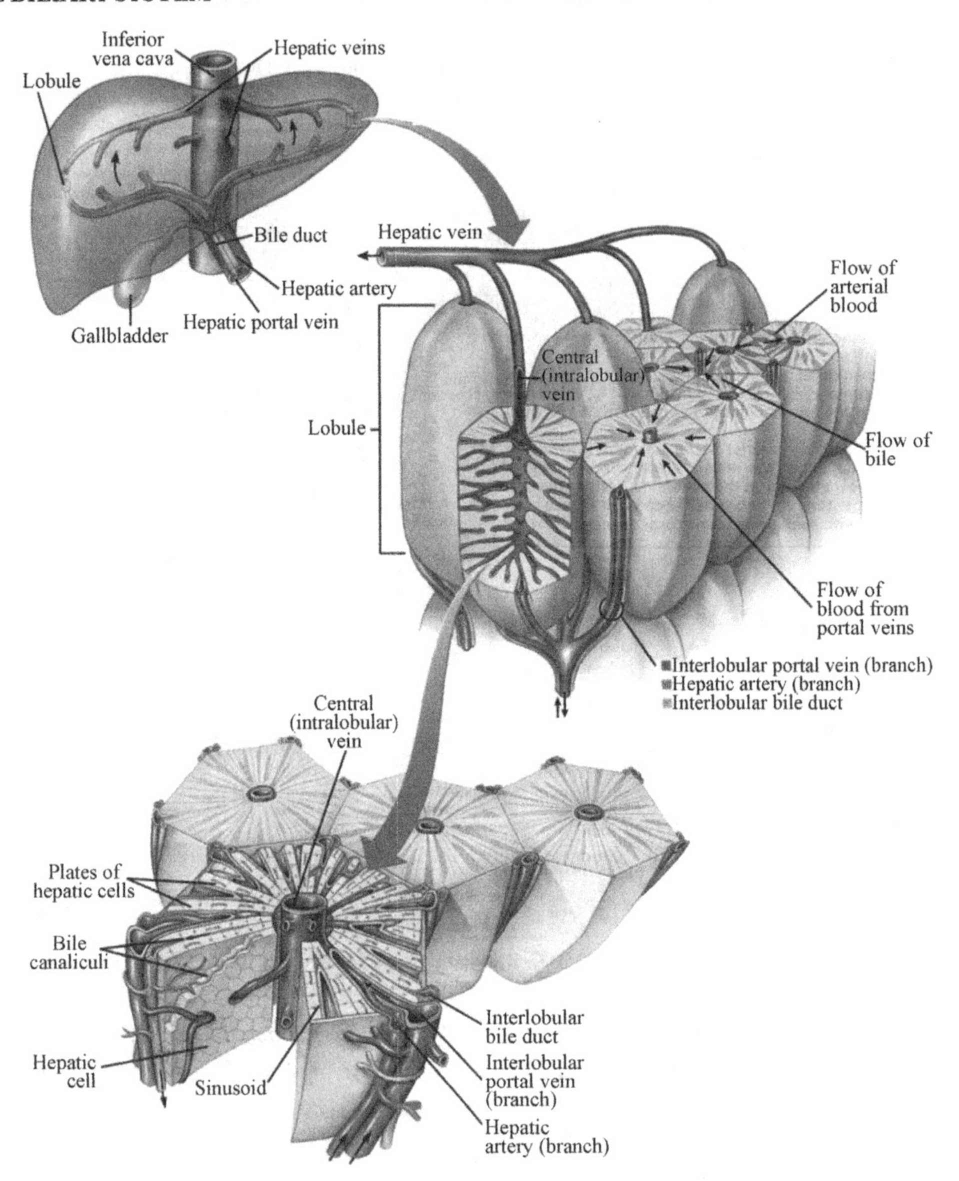

FIGURE 2.4: Microscopic structure of the liver. Top panel shows the location of liver lobules relative to the overall circulatory scheme of the liver. Middle and bottom panels show enlarged views of several lobules. Blood from the hepatic portal veins and hepatic arteries flows through sinusoids and thus past plates of hepatic cells toward a central vein in each lobule. Hepatocytes form bile, which flows through bile cannaliculi toward hepatic ducts that eventually drain the bile from the liver. Used with permission from *Anatomy & Physiology*. Editors: Thibodeau GA and Patton KT. 4th Edition. Mosby. St. Louis. 1999. p. 752.

hepatic arteries each divide into two arteries that supply the right anterior and posterior sections and the left medial and lateral sections, respectively. The middle hepatic artery arises from the right or left hepatic artery and supplies the quadrate lobe [9]. The hepatic artery blood flow provides oxygen and nutrients to the tissues of portal tracts, the liver capsule, and the walls of large vessels. In portal tracts, arterial branches form a capillary network arborized around bile ducts. Although arterial and portal blood appear to be well mixed before entering sinusoids, the direct supply of arterial blood to sinusoids by small branches of the hepatic artery remains unknown.

Hepatic Veins

There are three main hepatic veins (Figure 2.4). The middle and left veins unite before entering the vena cava in 65% to 85% of individuals. In 18% of individuals, there are two right hepatic veins draining into the vena cava. In another 23%, there is a separate middle or inferior right hepatic vein draining segments V or VI, respectively. A small amount of venous drainage from the liver surrounding the cava drains directly into the cava via small veins.

Lymphatic Drainage

The formation of hepatic lymph is poorly understood yet overproduction of hepatic lymph is responsible for the development of ascites in the setting of obstruction of sinusoidal blood flow. Lymphatic vessels can be identified in portal triads and these likely receive lymph from the space of Disse that exists between the fenestrated sinusoidal endothelial cells and the adjacent hepatocytes. A mechanism responsible for the countercurrent flow of lymph in spaces of Disse and flow of blood in sinusoids other than simple hydrostatic pressure remains unknown. A large volume of lymph (approximately more than 50% of all lymph) is produced in the liver. Lymphatic vessels from the gallbladder and cystic duct drain principally into the hepatic nodes via the cystic duct node, a constant lymph node located at the junction of the cystic duct and common hepatic duct [1–5]. Lymphatic vessels from the hepatic ducts and upper common bile duct drain into the hepatic lymph nodes, a chain of lymph nodes that follows the course of the hepatic artery to drain into the celiac lymph nodes. Lymph from the lower bile duct drains into the lower hepatic nodes and the upper pancreatic lymph nodes.

THE BILIARY TRACT AND GALLBLADDER

The path by which bile flows from the liver to the duodenum is as follows: bile secreted by the hepatocytes passes through canaliculi between hepatocytes to the center of the acinus, which is also the periphery of the lobule, to join small bile ducts (Figure 2.5). The small bile ducts within the right and left lobes of the liver join to form two larger ducts that emerge from the undersurface of

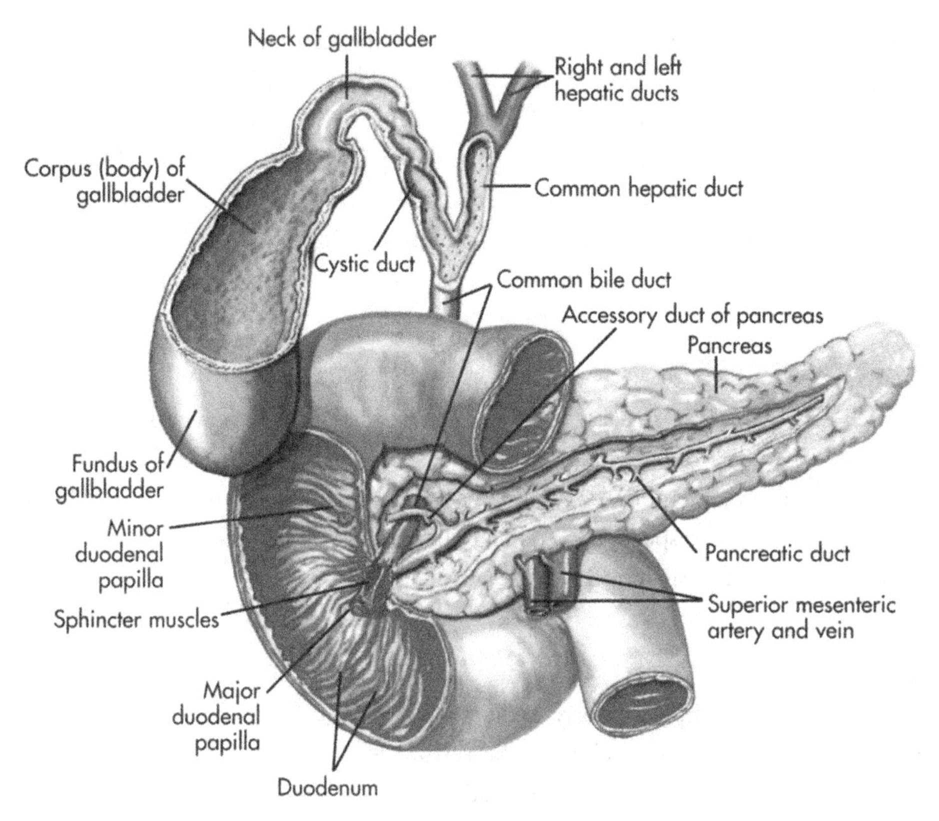

FIGURE 2.5: Common bile duct and its tributaries. Used with permission from *Anatomy & Physiology*. Editors: Thibodeau GA and Patton KT. 4th Edition. Mosby. St. Louis. 1999. p. 754.

the liver as the right and left hepatic ducts [1–4]. These two bile ducts immediately join to form one bile duct called the common hepatic duct. The common hepatic duct merges with the cystic duct from the gallbladder to form the common bile duct [10–15]. The latter opens into the descending part of duodenum called the major duodenal papilla (of Vater) [16]. This papilla is located 7 to 10 cm below the pyloric opening from the stomach. In general, the biliary tract is divided into three parts: the intrahepatic bile ducts, the extrahepatic bile ducts, and the gallbladder.

Intrahepatic Bile Ducts

The biliary drainage of the right and left lobes of liver is into the right and left intrahepatic bile ducts, respectively. The right hepatic duct is formed from the unification of the right posterior and right anterior segmental ducts. The left hepatic duct is formed by the unification of the three segmental ducts draining in the left side of the liver. The left hepatic duct crosses the base of segment IV in a horizontal direction to join the right hepatic duct and form the common hepatic duct.

Extrahepatic Bile Ducts

The right and left hepatic ducts often unite just outside of the liver parenchyma to form the common hepatic duct (Figure 2.5). The common hepatic duct is a segment of bile duct between the junction of the right and left hepatic ducts and the entrance of the cystic duct emanating from the gallbladder, and its length is variable. The common bile duct is formed by the unification of the cystic duct and the common hepatic duct. Its average length is approximately 8 cm, which can vary depending on the point of union of the cystic duct and the common hepatic duct. Generally, the diameter of the common bile duct varies from 4 to 7 mm. If its diameter exceeds 7.5 mm by imaging, the common bile duct is considered distended.

The relationship between the distal common bile duct and pancreatic duct is variable (Figure 2.6). In most instances (90%), the common bile duct and pancreatic duct join to form the common channel, which is less than 1.0 cm in length and is called the ampulla (meaning "jug"). In rare

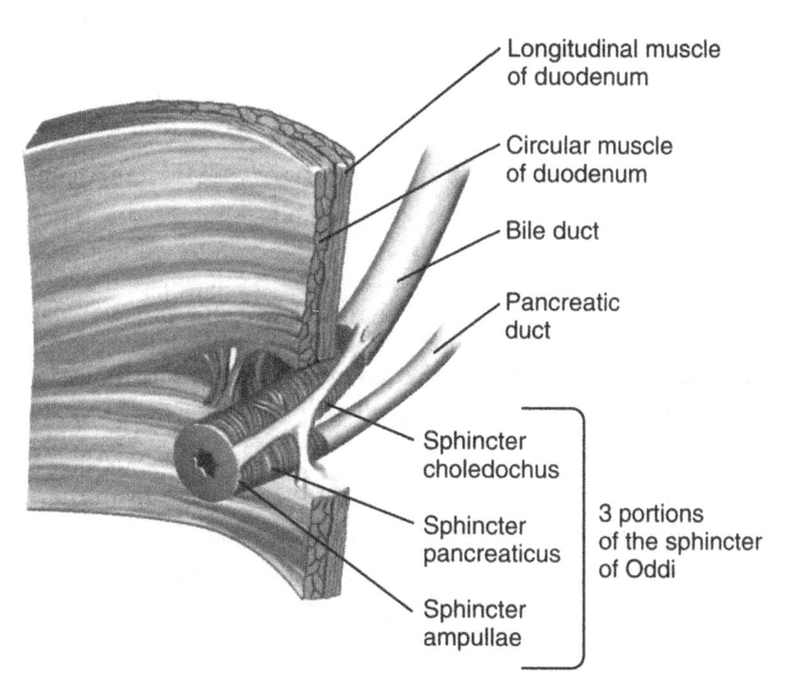

FIGURE 2.6: Anatomy of the sphincter of Oddi. This diagram shows the three portions of the sphincter of Oddi: the sphincter ampullae (surrounding the short common channel), the sphincter pancreaticus, and the sphincter choledochus (the largest portion). Used with permission from Elmunzer BJ and Elta GH. Biliary Tract Motor Function and Dysfunction in *Sleisenger and Fordtran's Gastrointestinal and Liver Disease*. Editors: Feldman M, Brandt L, and Friedman L. the 9th Edition. Elsevier Saunders. 2010; p. 1068.

situations (10%), these two structures may unite outside the duodenal wall to form a longer than 1.0 cm common channel, or alternatively the biliary and pancreatic ducts can drain separately into the duodenum [6, 17].

The sphincter of Oddi is usually considered to be composed of the lower portion of the common bile duct and the terminal portion of the pancreatic duct (Figure 2.6). The sphincter mechanism functions independently from the surrounding duodenal musculature and has separate sphincters for the distal bile duct, the pancreatic duct, and the ampulla. The entire sphincter mechanism is actually composed of four sphincters containing both circular and longitudinal smooth muscle fibers, i.e., the superior and the inferior sphincter choledochus, the sphincter pancreaticus, and the sphincter of the ampulla.

Gallbladder

The gallbladder is a pear-shaped organ located on the inferior surface of the liver at the junction of the right and left hepatic lobes, which typically hangs from the anterior inferior margin of the liver. It is found on the right side just deep to where the lateral margin of the rectus abdominis muscle crosses the costal margin of the rib cage. In general, the size of gallbladder varies between 7 and 10 cm in length and between 2.5 and 3.5 cm in width. The volume of a moderately distended gallbladder is approximately between 30 and 60 mL. Furthermore, the gallbladder's volume varies considerably, being large because of the storage of concentrated bile during fasting states and becoming small due to its postprandial emptying [18]. The gallbladder can be divided into four parts: the neck, body, infundibulum, and fundus (Figure 2.5). The neck of gallbladder connects the cystic duct in a cephalad and dorsal direction. The cystic duct often joins the lateral aspect of the supraduodenal portion of the common hepatic duct to form the common bile duct. The length of the cystic duct varies from 2 to 4 cm. The cystic duct may irregularly join the right hepatic duct or extend downward to connect the retroduodenal bile duct. The body is the central part of the gallbladder. The fundus projects downward beyond the inferior border of the liver. Hartmann's pouch is an asymmetrical bulge of the infundibulum close to the gallbladder's neck. Calot's triangle is formed by the common hepatic duct medially, the cystic duct laterally, and the cystic artery superiorly [19]. During cholecystectomy, a clear visualization of Calot's triangle is crucial with accurate identification of all structures within this triangle (Figure 2.3). In most cases, the cystic artery arises as a branch of the right hepatic artery within this triangle.

LIVER, GALLBLADDER, AND BILE DUCT CELL TYPES

Parenchymal Cells

Hepatocytes

The microscopic architecture of the liver has been divided into functional units called lobules based around central veins or acini based around portal triads (Figure 2.7). The hepatocytes of the portal

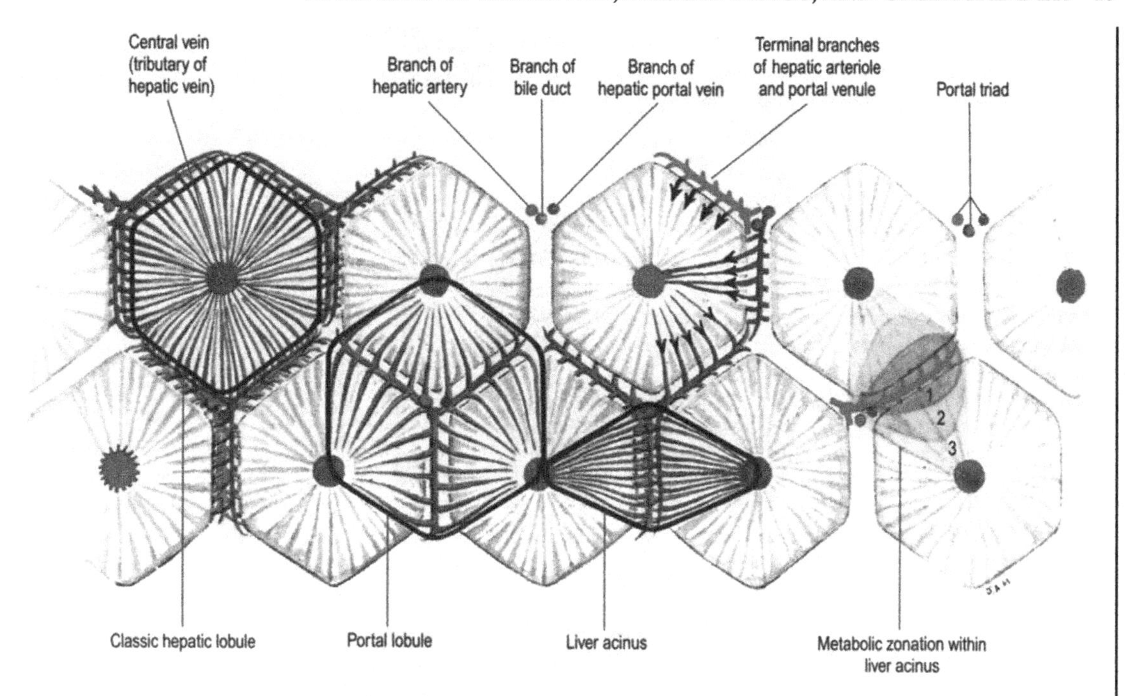

FIGURE 2.7: The histological organization of the liver. This diagram shows the principal types of subdivisions. Although the classic hepatic lobules are shown as regular hexagons, their real appearance is highly variable. The portal lobule, centered on the portal triad and biliary drainage is also shown. Used with permission from *Gray's Anatomy: The Anatomical Basis of Clinical Practice.* Editor-in-Chief: Standring S. 39th edition. Elsevier Churchill Livingstone, London, 2005. p. 1222.

based acinus are further subdivided into zone 1 hepatocytes that are closest to the portal triad and are the first to receive nutrient rich and well oxygenated blood, zone 3 hepatocytes which are most distal from the portal triads and zone 2 hepatocytes in between. The acinar structure has functional significance since zone 1 hepatocytes exhibit significant functional differences from zone 3 hepatocytes based on their respective roles in metabolism. The lobule is more of an anatomical concept with less functional significance. Between the portal triads bringing blood into the liver and the central veins are the hepatocytes that are arranged in irregular, branching, interconnected plates around the central vein (Figure 2.8). Hepatocytes are large polyhedral cells and account for approximately 70% of cells within the liver. Their cell sizes vary from 20 to 30 μm in diameter. Hepatocytes are polarized epithelial cells and their plasma membranes have three distinct domains: (i) the basolateral sinusoidal surface (about 37% of the cell surface) being in direct contact with plasma via the fenestrae of the specialized hepatic sinusoidal endothelial cells; (ii) the apical canalicular surface (about 13% of the cell surface) enclosing the entire bile canaliculus; and (iii) contiguous lateral surfaces (about half of the cell surface) adjacent to other hepatocytes. The sinusoidal and canalicular surfaces

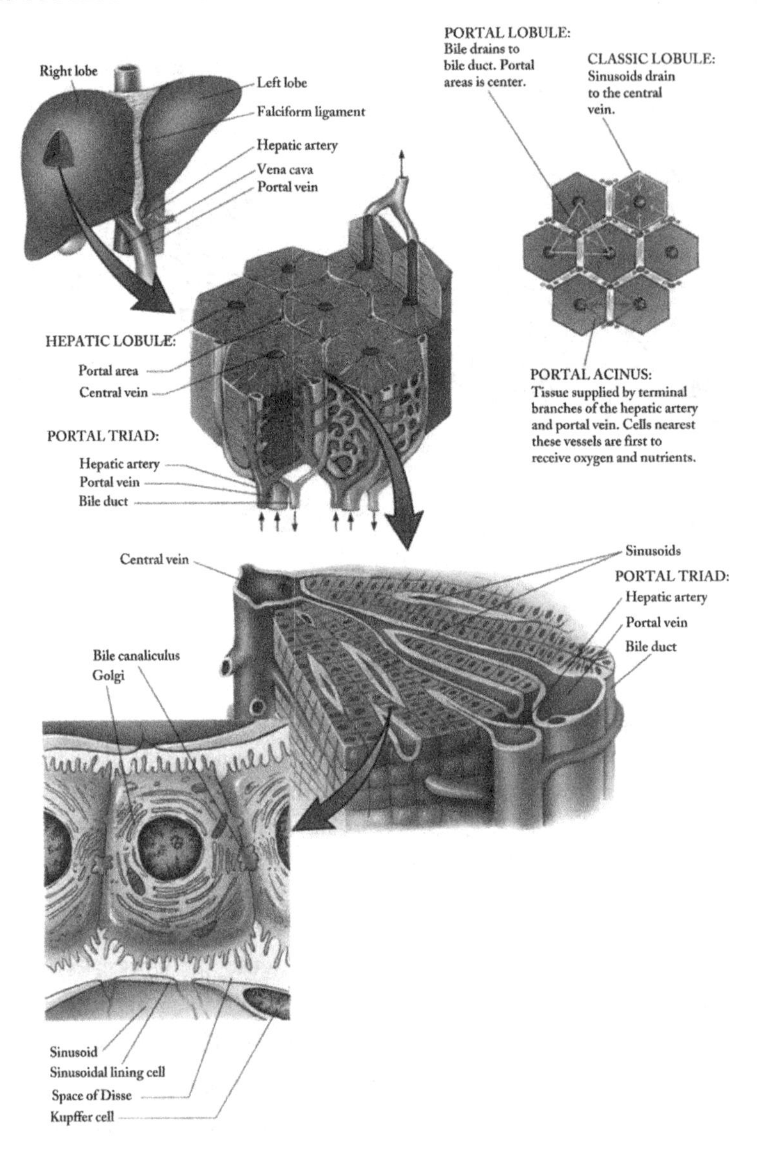

FIGURE 2.8: Microscopic anatomy of the liver. Left top panel shows schematic three-dimensional representation of a liver lobule. Right bottom panel shows enlarged view of a small part of a liver lobule. The directions of blood flow and bile flow are indicated by arrows; however, their directions are opposite. Used with permission from *Color Atlas of Histology*. Editors: Gartner LP and Hiatt JL. 3rd edition. Lippincott Williams & Wilkins, Philadelphia, 2000. p. 301.

contain a large number of microvilli, which significantly enlarge the surface area of these domains. The space between the endothelia and the sinusoidal villi is termed the space of Disse, which provides room for the bidirectional exchange of water and solutes between the plasma and hepatocytes at the sinusoidal surface. There are many transporter proteins located on the basolateral membrane for the molecular transfer of solutes, which promote facilitated diffusion or energy-consuming active transport. The canalicular domains of two adjacent hepatocytes are sealed at the periphery by tight junctions and form the bile canaliculus, which is the beginning of the biliary drainage system.

Sinusoidal Nonparenchymal Cells

Cells within the liver that are not hepatocytes are collectively called nonparenchymal cells. This is a diverse population serving a wide variety of metabolic, immunologic and structural functions.

Hepatic Sinusoidal Endothelial Cells

Instead of true capillaries, the liver has larger, endothelium-lined spaces called sinusoids, through which blood passes at a relatively low pressure (5–10 mm Hg). Different from capillary endothelial cells, these hepatic sinusoidal endothelial cells do not form intracellular junctions and simply overlap one another. These cells account for about 20% of total liver cells. Hepatic sinusoidal endothelial cells have fenestrae, or holes, in their flat, thin extensions to form sieve plates. Because of the presence of fenestrae and the absence of a basement membrane, this special arrangement allows plasma to enter the space of Disse and be in direct contact with the sinusoidal basolateral surfaces of hepatocytes. This greatly enhances bidirectional exchanges of liquids and solutes between the plasma and hepatocytes and it may serve as an important function in allowing large molecules such as lipoproteins and albumin to be in direct contact with the hepatocyte membranes. Diameters of the fenestrae are determined actively by the actin-containing components of the cytoskeleton in response to changes in the chemical environment. Therefore, the specialized endothelial lining of hepatic sinusoids serves as a selective barrier between the blood and the hepatocytes. In addition, hepatic sinusoidal endothelial cells can secrete prostaglandins, interleukins 1 and 6, tumor necrosis factor-α, interferon, and endothelin.

Kupffer Cells

Kupffer cells are the resident liver macrophages and they originate from bone marrow stem cells or monocytes. They are located within the sinusoidal lumen and have a direct contact with endothelial cells. These cells are highly active in cleaning up toxic or foreign substances and particulate matters derived from the intestines. They increase in number and activity in response to chemical, infectious, or immunologic injury to the liver. Kupffer cells contain ample lysosomes and can secrete a variety of noxious mediators for degrading substances taken up from the blood stream. Because of these functions, Kupffer cells are involved in host defense mechanisms and in the pathophysiological processes in some liver diseases.

Perisinusoidal Nonparenchymal Cells

Hepatic Stellate Cells

Hepatic stellate cells are located in the space of Disse between the endothelial cells and hepatocytes. The flat cytoplasmic extensions of quiescent hepatic stellate cells spread out parallel to the endothelial lining and contact several cells (Figure 2.9). Although hepatic stellate cells account for

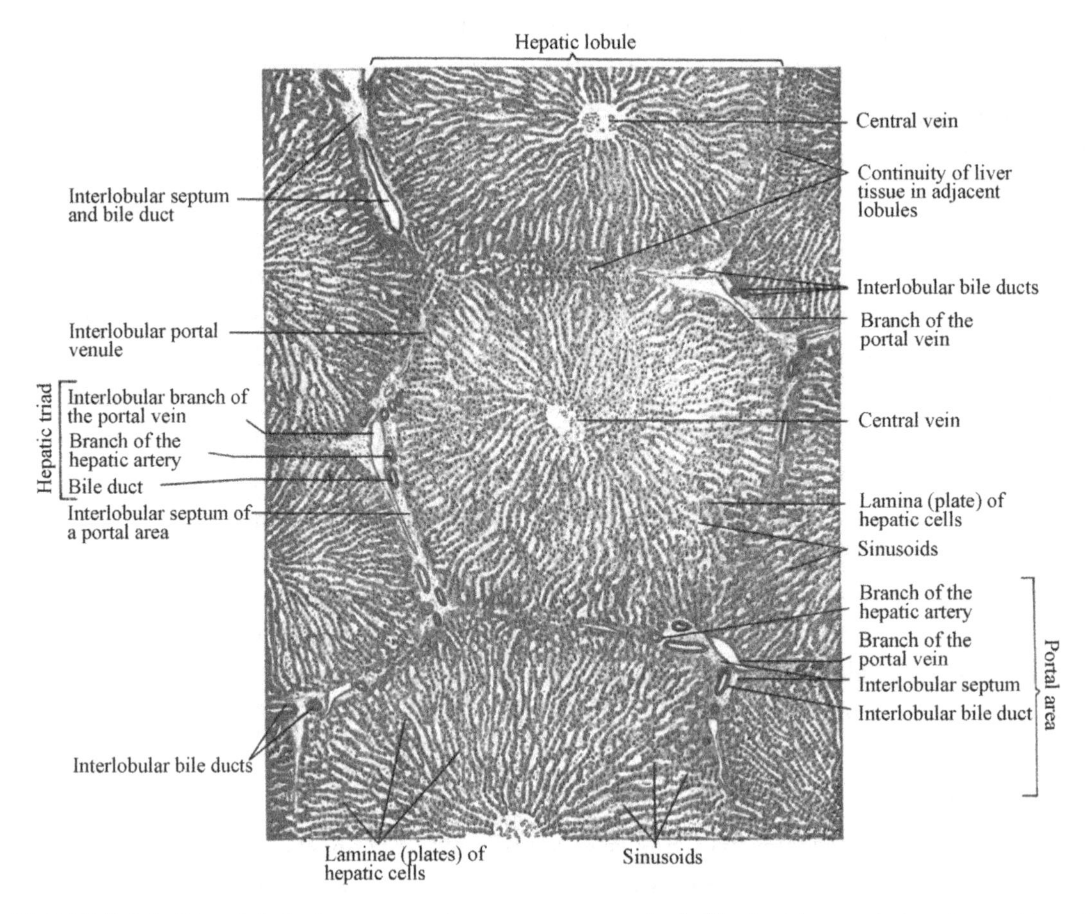

FIGURE 2.9: Liver lobule (panoramic view, transverse section). Used with permission from *Atlas of Human Histology*. Editor: Di Fiore MSH. 5th edition. Lea & Febiger, Philadelphia, 1981. p. 165.

about 5% of all liver cells, they are important sources of paracrine, autocrine, juxtacrine, and chemo attractant factors that maintain homeostasis in the microenvironment of the hepatic sinusoid. After chronic liver injury, these mesenchymal cells become activated, as evidenced by the loss of the retinoid containing droplets and up-regulation of the synthesis of extracellular matrix components such as collagen, proteoglycan, and adhesive glycoproteins. The activation of hepatic stellate cells is the central event in hepatic fibrosis. The overall number of hepatic stellate cells rises during fibrosis, because of a change in the balance between proliferation and apoptosis and the modulation of stellate cell activation and proliferation has become a focus of intense research activity to prevent the development of cirrhosis.

Pit Cells

Pit cells have the appearance of large lymphocytes and are natural killer cells of the liver. They are located primarily within the sinusoidal lumen and are adherent to the sinusoidal wall, often anchored with villous extensions. Pit cells are short-lived and are often replaced from extrahepatic sources. Similar to circulating natural killer cells, hepatic pit cells express OX-8 antigens and asialo-GMr1. However, they do not express the pan-T-cell marker, OX-19, which is expressed by circulating natural killer cells. Pit cells have tumor cell-killing activity in the liver and may play a role in eliminating virus-infected hepatocytes.

Gallbladder Cells

The gallbladder wall consists of five layers: the epithelium, lamina propria, smooth muscle, subserosal connective tissue, and serosa (Figure 2.10). The gallbladder lacks submucosa or muscularis mucosa. Most cells in the mucosa are simple columnar epithelial cells that are aligned in a single row and function mainly for absorption and secretion. The lamina propria contains nerve fibers, vessels, lymphatics, elastic fibers, loose connective tissue, and occasional mast cells and macrophages. The muscle layer is a loose arrangement of circular, longitudinal, and oblique fibers without well-developed layers. These muscles contract to excrete concentrated gallbladder bile into the intestinal lumen to help the digestion and absorption of dietary fat during the postprandial period (i.e., after food consumption). They relax for refilling hepatic bile by concentrating and storing it in the gallbladder during the interprandial period (i.e., in the fasting state). The subserosa is composed of a loose arrangement of fibroblasts, elastic and collagen fibers, vessels, nerves, lymphatics, and adipocytes. The gallbladder's outer coat is the visceral peritoneum. Rokitansky–Aschoff sinuses are invaginations of epithelium into the lamina propria, muscle, and subserosal connective tissue. These sinuses are found in approximately 40% of normal gallbladders and can be detected in almost all inflamed gallbladders.

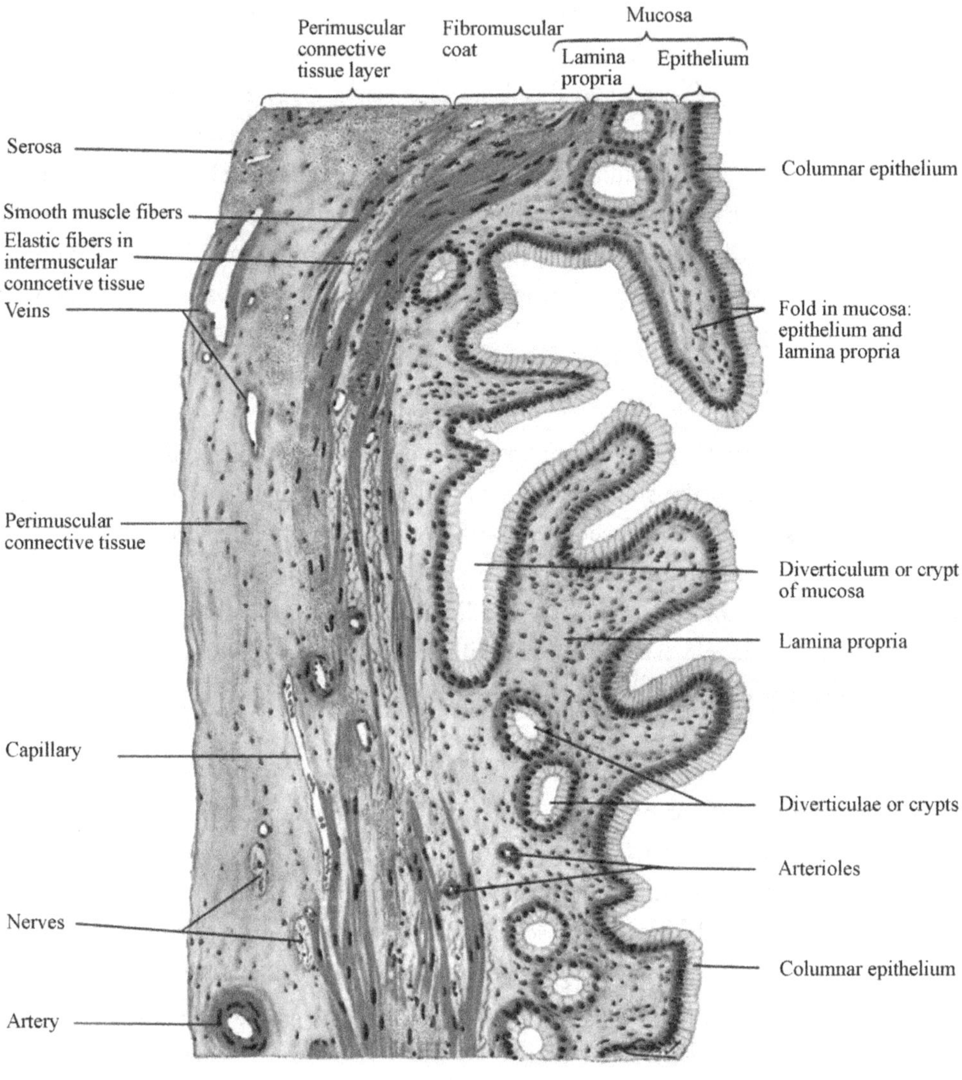

FIGURE 2.10: The histology of gallbladder. Used with permission from *Atlas of Human Histology*. Editor: Di Fiore MSH. 5th edition. Lea & Febiger, Philadelphia, 1981. p. 171.

Bile Duct Epithelial Cell

The bile ducts contain a columnar mucosa surrounded by a connective tissue layer. Bile duct epithelial cells, or cholangiocytes, are comprised of large and small subpopulations of cells, the cell sizes of which correlate approximately with the diameter of the bile ducts. Cholangiocytes are highly polarized, and the surface of epithelial cells is relatively flat, with basal nuclei and an absent or small nucleolus. Muscle fibers in the bile duct are sparse and discontinuous. The muscle fibers are usually longitudinal, although occasional circular fibers are found. The distal common bile duct begins to develop a more substantial muscle layer in the intraduodenal portion of this bile duct. At the sphincter of Oddi, such a structure becomes prominent and many bundles of longitudinal and circular fibers are noticeably identifiable. The lamina propria consists of collagen, elastic fibers, and vessels.

Bile ducts are not simple passive conduits for biliary drainage because they can play an active role in the absorption and secretion of biliary components and in the regulation of the extracellular matrix composition. It has been found that on the apical membrane of the cholangiocytes in the large bile ducts, bile acids are absorbed by an apical sodium-dependent bile acid transporter (ASBT). Bile acids may then exit at the basolateral membrane of cholangiocytes into the hepatic arterial circulation via the heteromeric organic solute transporter OSTα/β or an ATP-dependent transporter ABCC3. Cholangiocytes also contain a chloride channel that corresponds to the cystic fibrosis transmembrane regulator (CFTR) and a chloride–bicarbonate anion exchanger isoform 2 (AE2) for secretion of bicarbonate. In addition, secretin and somatostatin receptors are expressed in the large, but not the small, cholangiocytes, which may enable this type of cholangiocyte to regulate water and electrolyte secretion in response to secretin and somatostatin. Activation of apical purinergic receptor by ATP stimulates Ca^{2+} stores, thereby promoting Cl^{-} efflux from cholangiocytes. Aquaporin-1 on the apical and basolateral membranes constitutes water channels regulating hormone-regulated transport of water into bile by cholangiocytes. The purinergic receptor stimulates chloride ion efflux. Moreover, the sodium-dependent glucose transporter (SGLT1), an active glucose transporter on the apical (luminal) surface of cholangiocytes, and the glucose transporter 1 (GLUT1), a facilitative glucose transporter on the basolateral domain, are both responsible for glucose reabsorption from bile.

• • • •

CHAPTER 3
Physical Chemistry of Bile

CHEMICAL COMPOSITIONS OF BILE

Bile is a yellow, brownish, or olive-green liquid with a pH of 7.6–8.6. It is composed primarily of water, organic solutes, and inorganic electrolytes. In bile, cholesterol, phospholipids, and bile acids are three major lipid species (Figure 3.1), which account for approximate 99% of total lipids by weight. Bilirubin is a minor solute and represents less than 1% of biliary lipids. Bilirubin gives bile a yellow color.

Besides lipids, proteins and elements are also found in bile. Hepatic bile contains proteins at a concentration of approximately 0.2% of total mass by weight, about 30 times less than the protein concentration of plasma. Albumin is the most abundant protein in bile, followed by immunoglobulins G and M, apolipoproteins AI, AII, B, CI, and CII, transferring, and α_2-macroglobin. Other proteins that have been identified but not quantitated in bile include the epidermal growth factor, insulin, haptoglobin, cholecystokinin, lysosomal hydrolase, and amylase. Bile also contains a variety of other lipophilic substances, including vitamins, steroids, drugs, and other xenobiotics. Concentrations of various inorganic acids are similar to those in plasma, and the total mass of these acids in bile is approximately 9 g/L. These inorganic ions are responsible for most of the osmotic activity of bile. Elements detected in bile include sodium, phosphorus, potassium, calcium, copper, zinc, iron, manganese, molybdenum, magnesium, and strontium. Calcium ions can bind to bile acid micelles and may affect Ca^{2+} anion acid precipitation, which may also influence cholesterol precipitation. Table 3.1 shows the major chemical compositions of human hepatic and gallbladder biles.

Bile is partially an excretory product and partially a digestive secretion. The hepatic secretion of biliary cholesterol and its degradation product, bile acids, represents the major route for elimination of cholesterol from the liver, and eventually, from the body. After entering the intestinal lumen, bile acids play an important role in the emulsification of dietary lipids and the breakdown of large lipid globules into a suspension of droplets with sizes being about 1 μm in diameter. The tiny lipid droplets present a very large surface area, which allows pancreatic lipase to more rapidly accomplish digestion of dietary fat such as triglyceride. In addition, bile acids promote the intestinal absorption of cholesterol, fatty acids, fat-solvable vitamins (A, D, E, and K), and certain drugs.

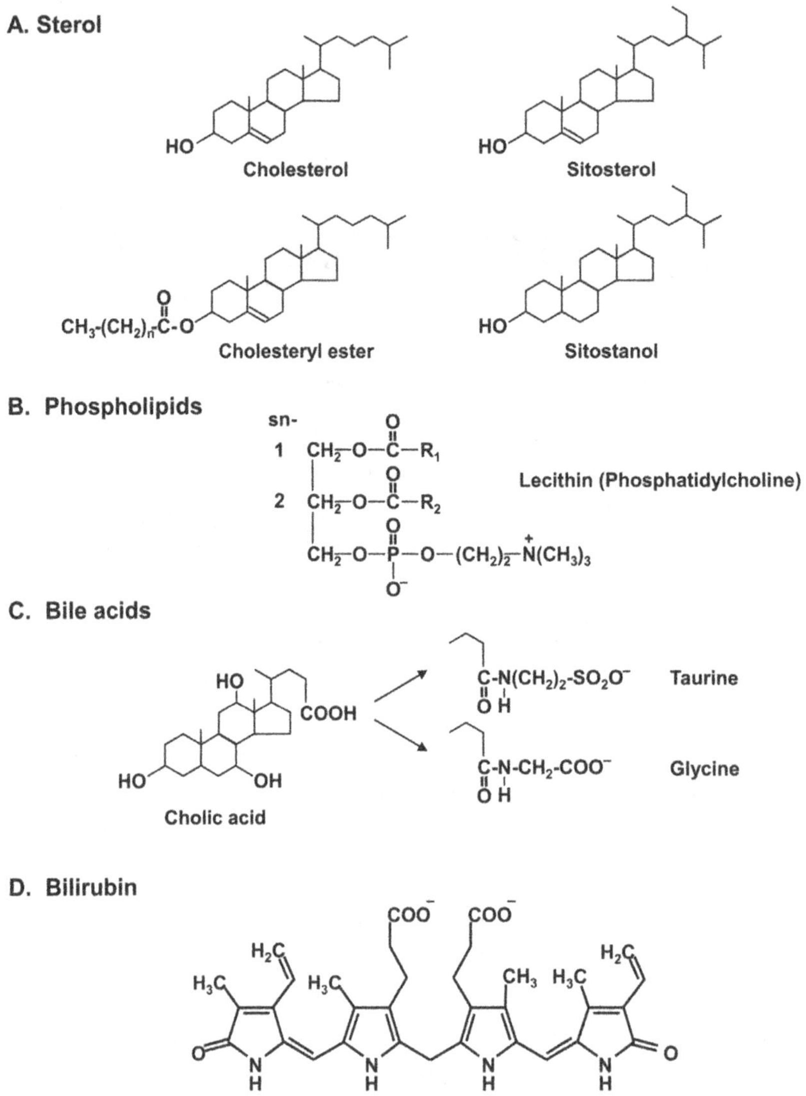

FIGURE 3.1: Chemical structures of major biliary lipids. (**A**) Cholesterol is one of the most abundant steroids in bile. Its hydroxyl group on the third carbon can react with the COOH group of a fatty acid molecule to form a cholesteryl ester. Plant sterols (e.g., β-sitosterol and β-sitostanol) are naturally occurring. Their chemical structures are very similar to cholesterol, but with structural modifications of the side chain. (**B**) Phospholipids are also derivatives of glycerol and contain a phosphate ester functional group and ionic charges, as illustrated for lecithin (phosphatidylcholine), which is the major phospholipid in human bile. In general, the sn-1 position of lecithin is esterified with a saturated fatty acid and the sn-2 position is esterified with an unsaturated fatty acid. (**C**) Bile acids are a family of closely related acidic sterols that are synthesized from cholesterol in the liver. The common bile acids, as represented by cholic acid that is the primary hepatic catabolic product of cholesterol, possess a steroid nucleus of four fused hydrocarbon rings with polar hydroxyl functions and an aliphatic side chain conjugated in amide linkage with taurine or glycine. (**D**) Bilirubin gives bile a yellow color. This figure shows the tetrapyrrole structure of bilirubin in its traditional open linear representation.

TABLE 3.1: Composition of hepatic and gallbladder biles in humans.

	HEPATIC BILE	GALLBLADDER BILE
Specific gravity	1.009–1.013	1.026–1.032
pH	7.1–8.5	5.5–7.7
Total solids, %	1–3.5	4–17
Total base, meq/L [a]	150–180	
Chloride, meq/L	75–110	15–30
Lipids, % bile acids + phospholipids + cholesterol		
Bile acids [b]	71.3	77.5 ± 4.7
Phospholipids	21.1	15.6 ± 4.8
Cholesterol	7.6	6.9 ± 2.0
Lipids, μmol/ml		
Bile acids		148.2 ± 25.1
Phospholipids		38.3 ± 6.5
Cholesterol		13.2 ± 3.1
Proteins, μg/ml		
Total protein		97.0 ± 12.9
Albumin	155–1485 (405)	
Transferrin	11.4–160 (36.3)	
α_2-Macroglobin	2.7–100 (13.5)	
Immunoglobulin G	32–480 (88.8)	
Immunoglobulin M	2.2–60 (19.6)	
Apoprotein AI	2.9 ± 0.5	19.1 ± 2.2
Apoprotein AII	1.5 ± 0.4	10.4 ± 1.1

TABLE 3.1: (*continued*)

	HEPATIC BILE	GALLBLADDER BILE
Apoprotein CI	12.4 ± 5.5	7.8 ± 1.4
Apoprotein CD	3.4 ± 1.1	3.9 ± 0.8
Apoprotein B	10.6 ± 5.2	38.5 ± 4.7
Elements, mM		
Ca		7.38 ± 2.92
Cu ($\times 10^2$)		9.58 ± 5.43
Fe ($\times 10^2$)		1.59 ± 1.32
K		12.68 ± 3.49
Mg		6.91 ± 2.23
Mn ($\times 10^2$)		1.18 ± 2.13
Mo		2.09 ± 1.04
Na		210.14 ± 12.13
P		54.18 ± 15.27
Sr		0.10 ± 0.10
Zn ($\times 10^2$)		1.77 ± 0.63

Values are means ± SE or range (means).

[a] Total base in gallbladder bile must be similar to that in hepatic bile. A variable component of total base is bicarbonate, which can be as high as 60 meq/L in hepatic bile but is usually quite low (1–5 meq/L) in gallbladder bile.

[b] The major bile acids in human bile are the primary bile acids, cholic acid and chenodeoxvcholic acid, which account for ~40 % each of bile acids. The secondary bile acids, deoxycholic acid (~20%) and lithocholic acid (~1%), account for the rest. Bile acids are conjugated mainly with glycine (~60%) and with taurine (~40%).

Used with permission from Cabral DJ and Small DM. Physical Chemistry of Bile in *Handbook of Physiology*, Volume III. Editors: Schultz SG, Forte JG, and Rauner BB. 1st edition. Waverly Press, New York, 1989. p. 651.

BILIARY CHOLESTEROL

Figure 3.2 shows the basic chemical structure of steroids that have a nucleus containing the four-ringed carbon skeleton of cyclopentenophenanthrene and the numbering of the carbon atoms in steroids.

The basic structure of the cholesterol molecule includes (i) the perhydrocyclopentenophenanthrene nucleus with its four fused rings, (ii) a single hydroxyl group at C-3, (iii) a double bond between C-5 and C-6, (iv) an eight-membered branched hydrocarbon chain attached to carbon

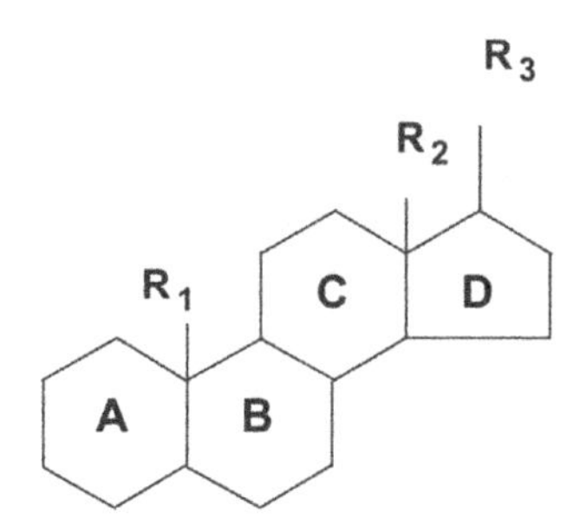

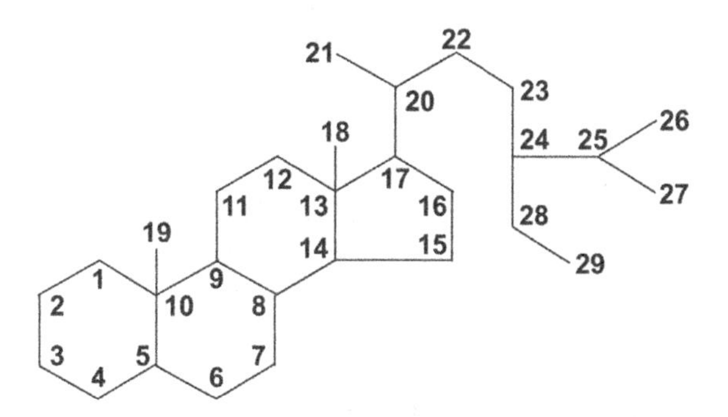

FIGURE 3.2: Many substances present in animals and plants, including those with diverse biological functions such as bile acids, are structurally related to cholesterol. All these substances have a nucleus containing the four-ringed carbon skeleton of cyclopentenophenanthrene and are known as steroids. The general structural formula for the steroids, including the designation of the four rings, is shown in the top panel. R_1 and R_2 are usually methyl groups and R_3 is usually a side-chain. The numbering of the carbon atoms in steroids is shown in the structural formula (bottom panel). If one or more carbon atoms is not present (e.g., C-28 and C-29 are not present in cholesterol), the numbering of the remaining carbons is unchanged. If one of the two methyl groups attached to C-25 is substituted, it is denoted C-26; e.g., 26-hydroxycholesterol (not 27-hydroxycholesterol).

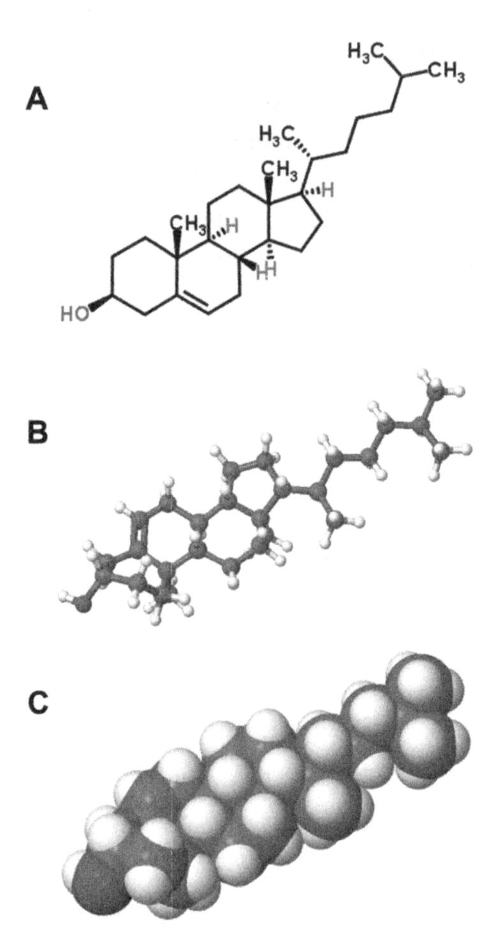

FIGURE 3.3: Molecular structure of cholesterol. The standard chemical formula (top panel), the perspective formula (middle panel), and the space-filling model (bottom panel) are shown.

17 in the D ring, and (v) a methyl group (carbon 19) attached to carbon 10, and a second methyl group (carbon 18) attached to carbon 13. Furthermore, in the esterified form, a long-chain fatty acid (usually linoleic acid) is attached by ester linkage to the hydroxyl group at C-3 of the A ring (Figure 3.3). Cholesterol is present solely in the unesterified form, which accounts for up to 95% of the total sterols by weight in bile. The remaining 5% of the sterols are cholesterol precursors and dietary sterols from plant, animal, and shellfish sources. These non-cholesterol sterols can be found in normal human bile, and their pattern and proportions are broad and highly dependent on diet. In general, on a regular (nonshellfish) diet, the concentrations of non-cholesterol sterols are less than 5% in bile, and their proportions are cholestanol (1.5%), sitosterol (1.2%), campesterol (0.7%),

lathosterol (0.6%), 24-methylene cholesterol (0.1%), stigmasterol (0.1%), brassicasterol (0.1%), and isofucosterol (0.03%). If a high shellfish diet is consumed, shellfish sterols would be increased in bile and could consist of 5–10% of total sterols by weight. Of note is that the concentrations of cholesteryl esters are negligible in human bile.

BILIARY BILE ACIDS

Bile acids make up approximately two thirds of the solute mass of normal human bile by weight and are a family of closely related acidic sterols that are synthesized from cholesterol in the liver. The common bile acids possess a steroid nucleus of four fused hydrocarbon rings with polar hydroxyl functions and an aliphatic side chain conjugated in amide linkage with glycine or taurine (Figure 3.1). In general, the glycine conjugate is more hydrophobic than the taurine conjugate. Because the ionized carboxylate or sulfonate group on the side chain renders bile acids to be water soluble, they are classified as soluble amphiphiles. The hydrophilic (polar) areas of bile acids are the hydroxyl groups and conjugation side chain of either glycine or taurine and their hydrophobic (nonpolar) area is the ringed steroid nucleus. In human bile, more than 95% of bile acids are 5β C_{24} hydroxylated acidic steroids, amide-linked to taurine or glycine in an approximate ratio of 1:3. The primary bile acids are hepatic catabolic products of cholesterol and are composed of cholic acid (a trihydroxy bile acid) and chenodeoxycholic acid (a dihydroxy bile acid). Figure 3.4 shows the chemical structure of cholic acid. Whereas the secondary bile acids are derived from the primary bile acid species by intestinal bacteria in the ileum and colon, being composed of deoxycholic acid, ursodeoxycholic acid, and lithocholic acid (Figure 3.5). The 7α-dehydroxylation of the primary bile acids is the most important reaction to produce deoxycholic acid and lithocholic acid from cholic acid and chenodeoxycholic acid, respectively. Another important secondary reaction is the 7α-dehydrogenation of chenodeoxycholic acid to form 7α-oxo-lithocholate. This bile acid does not accumulate in bile, but is metabolized to a "tertiary" bile acid by hepatic or bacterial reduction to form chenodeoxycholic acid (mainly in the liver) or its 7β-epimer, ursodeoxycholic acid (primarily by colonic bacteria).

BILIARY PHOSPHOLIPIDS

Phospholipids are derivatives of glycerol and contain a phosphate ester functional group and ionic charges. The phospholipids consist of 15–25% of total lipids by weight in bile. The major phospholipids are lecithins (phosphatidylcholines) in human bile, accounting for more than 95% of total phospholipids by weight. The remainder is composed of cephalins (phosphatidylethanolamines) and a trace amount of sphingomyelin. As a major phospholipid in bile, lecithins are insoluble, swelling amphiphiles with hydrophilic, zwitterionic phosphocholine head groups and hydrophobic

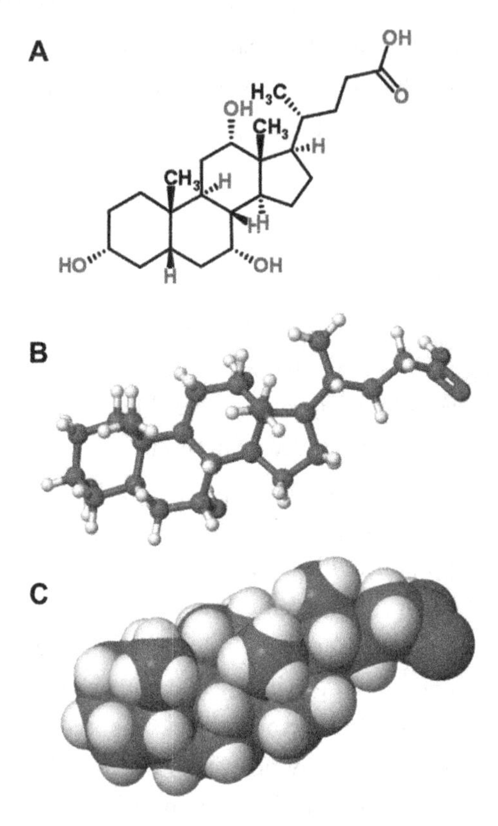

FIGURE 3.4: Molecular structure of cholic acid. The standard chemical formula (top panel), the perspective formula (middle panel), and the space-filling model (bottom panel) are shown. Although cholesterol and bile acids appear similar from their chemical formulae, the three-dimensional views reveal significant differences with these major changes: (i) hydroxylation at C-7; (ii) modified β-oxidation of the side chain that results in shortening of the C_8 isooctane side chain to a C_5 isopentanoic side chain; (iii) epimerization of the C-3 hydroxy group; and (iv) reduction of the double bond to give a 5β-bile acid in which the A/B ring junction is in the *cis* configuration. Saturation of the 5–6 double bond with addition of a 5β-hydrogen introduces a bend in the previously planar sterol nucleus. The addition of hydroxyl and side-chain carboxyl groups markedly increases the water solubility of cholic acid compared with cholesterol. Finally, the asymmetrically orientated hydroxyl groups on the bile acid molecule render it strongly amphophilic, with polar alpha and nonpolar beta surfaces.

tails with two long fatty acyl chains. Lecithins are synthesized from diacylglycerol predominantly in the endoplasmic reticulum of the liver by way of the cytidine diphosphate-choline pathway. Biliary lecithins are derived from the least hydrophobic hepatic lecithins and typically contain a saturated C_{16} acyl chain in the sn-1 position and an unsaturated C_{18} or C_{20} acyl chain in the sn-2 position (Figure 3.1). Furthermore, similar to all naturally occurring phospholipids, biliary lecithins are com-

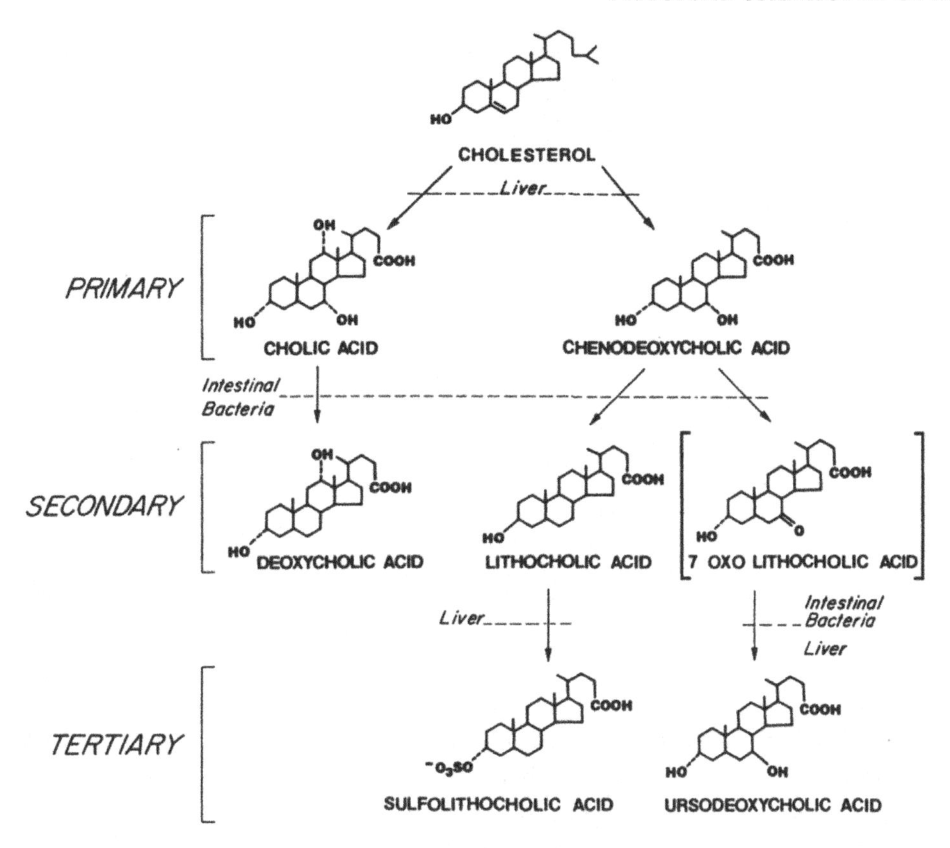

FIGURE 3.5: Major primary, secondary, and tertiary bile acids of humans with sites of synthesis and metabolism. Used with permission from Carey MC and Cahalane MJ. Enterohepatic Circulation in *The Liver: Biology and Pathobiology*. Editors: Arias IM, Jakoby WB, Popper H, Schachter D, and Shafritz DA. 2nd edition. Raven Press, New York, 1988. p. 576.

plex mixtures of molecular species (Figure 3.6). The sn-1 position is esterified by the saturated fatty acyl chains 16:0 (~75%) and 18:0 (less than 20%), with small amounts of monounsaturated sn-1 16:1 or 18:1 comprising the remainder. The sn-2 position is esterified by unsaturated fatty acyl species, with 18:2, 18:1 and 20:4 fatty acids predominating. The major molecular species of lecithins are 16:0–18:2 (40–60%), 16:0–18:1 (5–25%), 18:0–18:2 (1–16%) and 16:0–20:4 (1–10%) in human bile. Although there is a large variation in hepatic output of biliary bile acids, the proportion of lecithins to other phospholipid classes in bile is essentially constant.

BILE PIGMENTS

The principal bile pigment is conjugated bilirubins that are the fourth significant group of organic compounds found in bile. Chemically, the bile pigments are tetrapyrroles, which are derived from

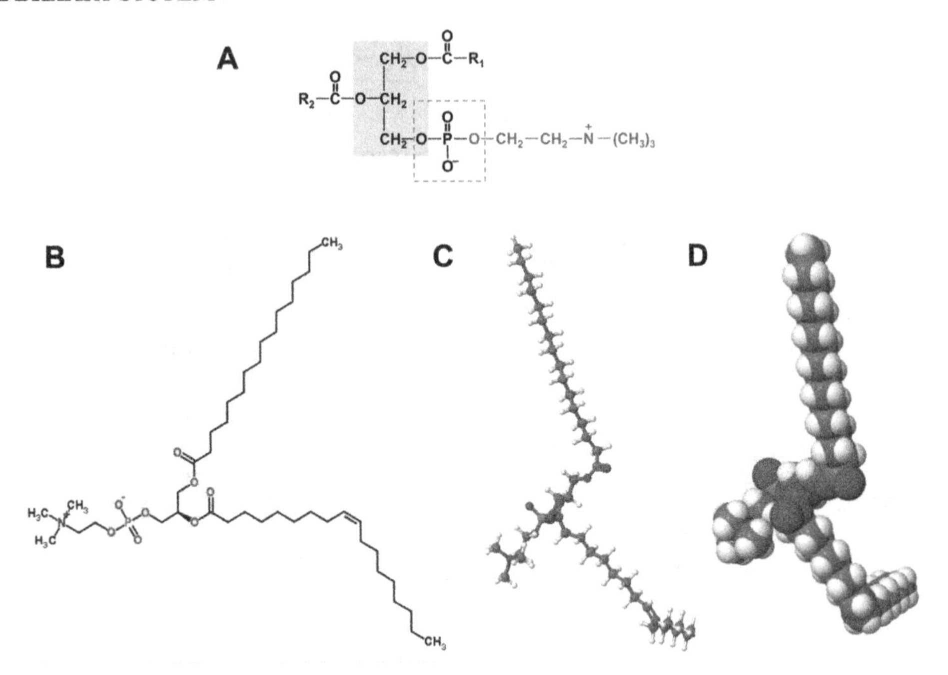

FIGURE 3.6: Phospholipids are polar, ionic lipids composed of 1,2-diacylglycerol (highlighted by a yellow rectangle) and a phosphodiester bridge (circled by a green box) that links the glycerol backbone to some base, usually a nitrogenous one, such as choline (highlighted in red color), serine, or ethanolamine. The standard chemical formula (top panel), the perspective formula (middle panel), and the space-filling model (bottom panel) for phosphatidylcholine (lecithin) are shown. Phosphatidylcholine is a major phospholipid in human bile and contains mostly palmitic acid (16:0) or stearic acid (18:0) in the sn-1 position and primarily unsaturated 18-carbon fatty acids oleic, linoleic, or linolenic in the sn-2 position.

porphyrins. The molecular structure of bilirubin is shown in Figure 3.7. Bilirubin is insoluble in water, but it becomes soluble after conjugation to glucuronic acid in the liver. Bilirubin is normally present in bile at a concentration of about 0.2 mM (0.2 g/L) and thus represents less than 1% of biliary solids by weight. Because bilirubin is secreted as the soluble salt bilirubin glucuronide, it is not found within the micelles. Approximately 80% of bilirubin is secreted as diglucuronide, 18% is secreted as monoglucuronide, and less than 2% is unconjugated. Small amounts of other related pigments are also present.

The phagocytosis of aged red blood cells releases iron, globin, and bilirubins derived from heme. The iron and globin are recycled, and some of the bilirubins are converted to conjugated bilirubins that are secreted into bile. Most of the bilirubins in bile are ultimately metabolized in the small intestine by bacteria and eliminated in feces. One of its breakdown products is stercobilin that gives feces a normal brown color.

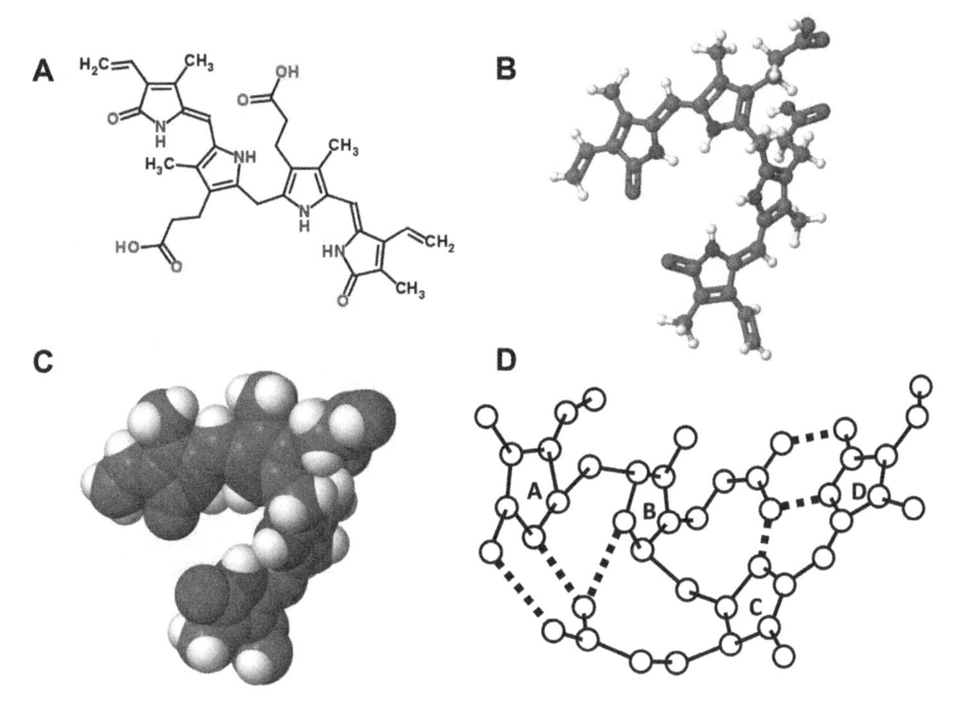

FIGURE 3.7: Molecular structure of bilirubin. (**A**) The standard chemical formula, (**B**) the perspective formula, and (**C**) the space-filling model are shown. (**D**) The tetrapyrrole structure of bilirubin in its fully internally bonded "ridge tile" configuration. The dashed lines represent the six internal hydrogen bonds between protonated carboxyl group and two amino nitrogens of the opposite dipyrrolic unit. In bile, the internal H-bonds are broken by conjugation of the propionate groups with glucuronic acid, xylose, glucose, or taurine.

• • • •

CHAPTER 4

Hepatic Cholesterol Metabolism

PHYSICAL CHEMISTRY OF CHOLESTEROL

Cholesterol is an essential component of mammalian cell membranes and is widely distributed in "free" (unesterified) and esterified forms (Figure 4.1). In its free form, the chemical structure of the cholesterol molecule includes the cholestene nucleus with a double bond at the C-5 and C-6 positions and a hydroxyl group on the third carbon (Figure 4.1). Furthermore, the angular methyl groups at C-10 and C-13, the hydrogen atom at C-8 and the side-chain at C-17 are in β configuration. The hydrogen atoms at C-9 and C-14 are in α configuration. In the plasma, approximately one third of cholesterol is in the free form and the remaining two thirds exist as cholesteryl esters [20–25]. Of note is that in bile, approximately 95% of the cholesterol molecule is in the free (unesterified) form and the remaining 5% of the sterols are cholesterol precursors and dietary sterols [26–28]. Moreover, the concentrations of cholesteryl esters are negligible in human bile. The solubility of cholesterol is very low in water, approximately 4.7 mM at 25°C. Furthermore, when one fatty acid attaches to the cholesterol molecule, its residue increases the hydrophobicity of cholesterol. The actual cholesterol concentration in plasma of a healthy individual is usually between 120 and 200 mg/dL. Such a high concentration of cholesterol can be present in the blood because plasma lipoproteins (mainly LDL, HDL, and VLDL) carry large amounts of cholesterol, regardless of whether the cholesterol molecule is in a free or an esterified form [29–35]. Also, cholesterol is abundant in human bile, with normal concentrations being approximately 390 mg/dL in the gallbladder. Bile acids, which are metabolites of cholesterol, can form simple and mixed micelles in bile, which can aid in solubilizing cholesterol in bile [36–41]. Furthermore, the vesicles that are composed primarily of phospholipids greatly promote the solubilization of cholesterol in bile [42–48].

FEATURES OF CHOLESTEROL BALANCE IN THE BODY

Cholesterol is the major sterol in humans and experimental animals. Besides being an important component of virtually all cell membranes, cholesterol is the precursor of various steroid hormones including the sex hormones (estrogen, testosterone, and progesterone) and corticosteroids (corticosterone, cortisol, cortisone, and aldosterone) [30, 49–51]. Figure 4.2 depicts the general features of cholesterol balance across the body. Because almost all of the cells in the major tissues need a

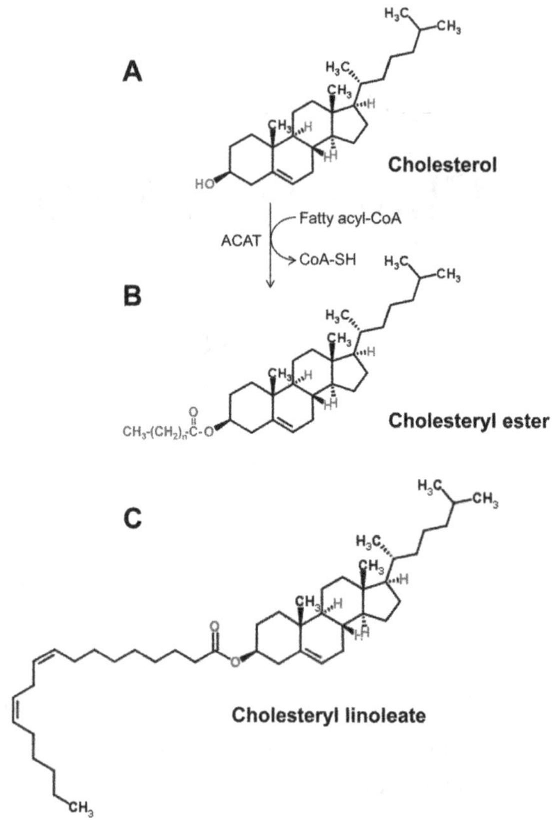

FIGURE 4.1: Chemical structures of (**A**) unesterified cholesterol (3β-cholest-5-en-3-ol); (**B**) the ACAT reaction, producing cholesteryl ester by acyl CoA:cholesterol acyl transferase (ACAT), and (**C**) cholesteryl linoleate (3β-cholest-5-en-3-yl (9Z,12Z)-octadeca-9,12-dienoate) where in the esterified form, a long-chain fatty acid (usually linoleic acid) is attached by ester linkage to the hydroxyl group at C-3 of the A ring.

continuous supply of cholesterol, a complex series of transport, biosynthetic, and regulatory mechanisms have evolved in the body [20, 52]. Furthermore, cholesterol can be obtained from both the intestinal absorption of dietary cholesterol and synthesized *de novo* from acetyl CoA within the body. Because human tissues do not possess enzymes that are able to degrade the ring structure of cholesterol, this sterol cannot be metabolized to CO_2 and water. So, the excess amounts of cholesterol must be metabolized and/or excreted to prevent a potentially hazardous accumulation of cholesterol in the body. This task is achieved in the body by modifying certain substituent groups on the hydrocarbon tail or on the ring structure of the cholesterol molecule. As a result, cholesterol is

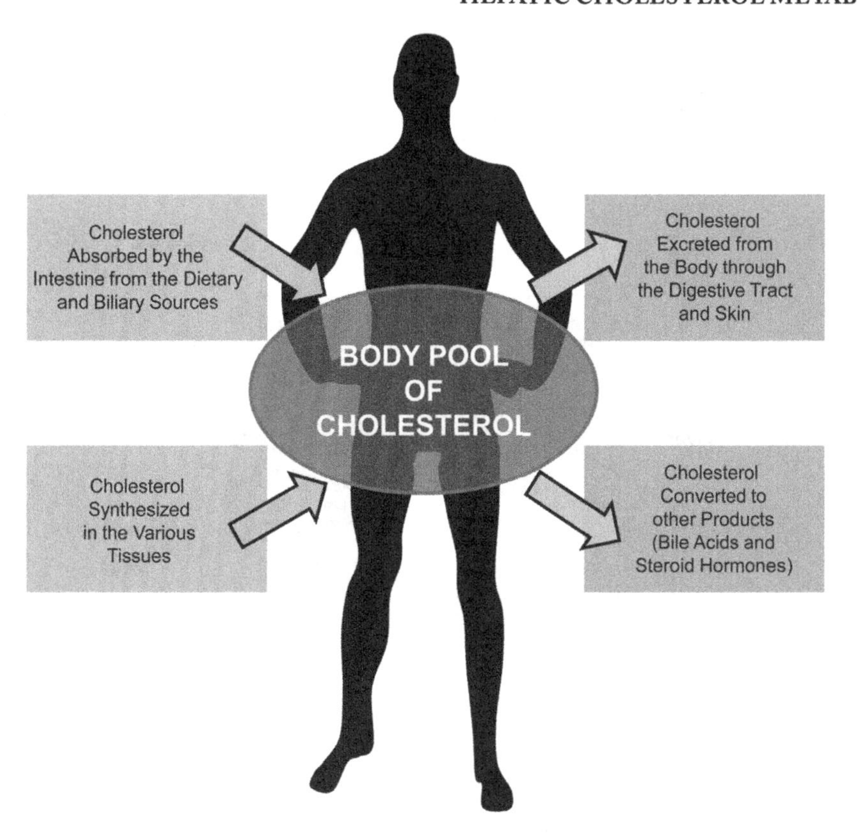

FIGURE 4.2: The general feature of cholesterol balance across the body. There are only two sources for cholesterol in the body: (i) intestinal absorption of cholesterol from dietary and biliary sources; and (ii) cholesterol biosynthesis in the various tissues. Likewise, there are two major pathways for the excretion of cholesterol from the body: (i) the excretion of cholesterol from the body through the gastrointestinal tract and skin; and (ii) conversion to various products, such as bile acids and steroid hormones. Because total input of cholesterol into the body must equal total output in the steady state, the body pool of cholesterol can be fundamentally kept constant. As a result, it prevents a potential accumulation of cholesterol in the body. Of note is that in children and growing animals, there is necessarily a greater input of cholesterol into the body than output since there is a net accumulation of cholesterol for keeping body weight gain.

often excreted from the body either as the unaltered molecule (i.e., in both free and esterified forms) or after biochemical modification to other sterol products such as bile acids and steroid hormones.

Figure 4.3 illustrates the pathways for the net flow of cholesterol through the major tissue compartments of the human and experimental animals. The body pool of cholesterol in the adult remains essentially constant [53–55]. New cholesterol can be added to the body pool from only two

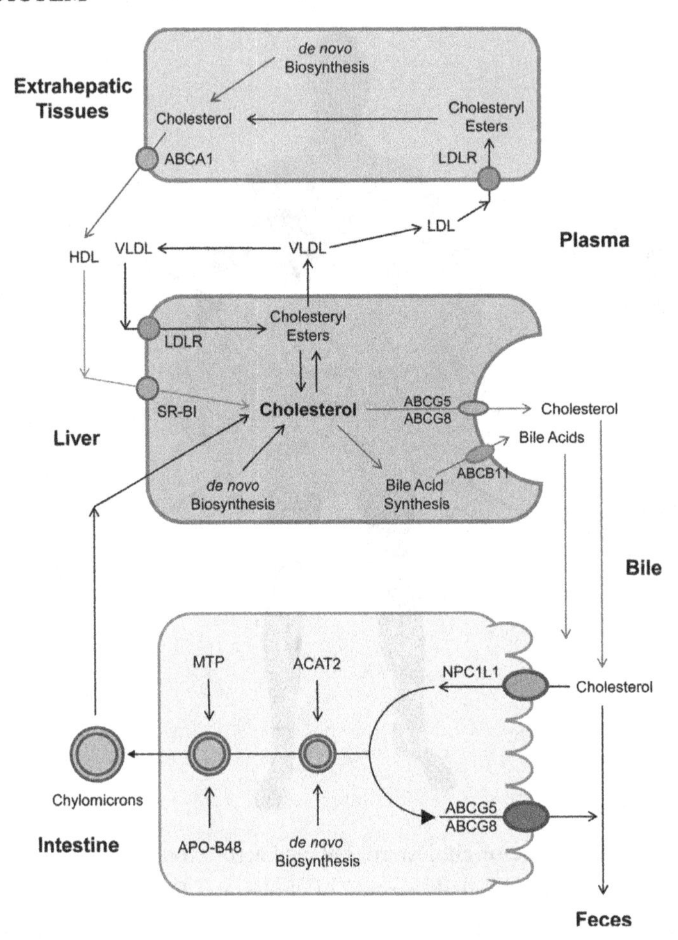

FIGURE 4.3: Pathways for the net flow of cholesterol through the major tissue compartments of the human and experimental animals. This diagram illustrates the major pathways for the net flow of cholesterol from the endoplasmic reticulum to the plasma membrane of the cells of the extrahepatic tissues (brown arrows), through the plasma space to the liver (green arrows), and finally, from the liver into the intestine and feces (red arrows). The specific proteins that may be involved in these pathways are shown in *white boxes*. Abbreviations: ABC, ATP-binding cassette (transporter); ACAT, acyl-coenzyme A:cholesterol acyltransferase; BA, bile acid; C, cholesterol; CE, cholesteryl ester; CM, chylomicron; LCAT, lecithin:cholesterol acyltransferase; LDLR, low-density lipoprotein receptor; NPC1L1, Niemann–Pick C1-like 1 protein.

sources: the absorbed cholesterol from dietary sources across the mucosa of gastrointestinal tract and the synthesized cholesterol in a variety of different tissues within the body. The availability of dietary cholesterol to the body varies enormously in different animal species and even in the same species, including humans, and the consumed amounts of dietary cholesterol also change markedly from day to day [53–64]. In addition, bile cholesterol is reabsorbed by the small intestine, which provides about two thirds of the total amount of cholesterol originating from the intestine every day [65]. Moreover, the total amount of cholesterol from the intestine to the body is dependent on the absorption efficiency of intestinal cholesterol and the amount of cholesterol that is consumed daily. Recent human and animal studies have revealed the ATP-binding cassette (ABC) transporters ABCG5 and ABCG8 are apical sterol export pumps promoting active efflux of cholesterol and plant sterols from the enterocytes back into the intestinal lumen for fecal excretion [66–76]. The Niemann–Pick C1 like 1 (NPC1L1) protein is also expressed at the apical membrane of enterocytes and plays a crucial role in the ezetimibe-sensitive cholesterol absorption pathway [77–81]. It has been found that cholesterol absorption is a multistep process, which is regulated by multiple genes at the enterocyte level, and the efficiency of cholesterol absorption could be determined by the net effect between influx and efflux of intraluminal cholesterol molecules crossing the brush border membrane of the enterocyte [82]. The 3-hydroxy-3-methylglutaryl CoA (HMG-CoA) reductase is the rate-limiting enzyme for cholesterol biosynthesis in the body [83–88]. Cholesterol that is synthesized *de novo* from acetyl CoA in various different tissues such as the liver and intestine is the second major source to the body [89–96]. Therefore, the sum of these two processes constitutes the total input of cholesterol into the body pool each day.

In humans and experimental animals, there are two major pathways for the removal of cholesterol from the body. The hepatic secretion of cholesterol into bile through the canalicular membrane of the hepatocyte is of greatest quantitative importance. In addition, the cholesterol molecule is first metabolized to other products such as bile acids, adrenocorticosteroids or testosterone, which, in turn, are excreted from the body through the gastrointestinal tract or urine. Of these various routes, the sterol efflux transporters ABCG5 and ABCG8 on the canalicular membrane of hepatocyte play a crucial role in the regulation of hepatic cholesterol secretion [72, 97–100], and ABCB11, a bile acid export pump, is responsible for hepatic secretion of biliary bile acids [101]. These transporters in the liver could have an important effect on regulating excretion of excess cholesterol from the body, either as free cholesterol or as its metabolic products—bile acids. In addition, the unmodified cholesterol molecules may be lost directly from the body pool. This takes place through the sloughing of oily secretions and cells from the skin, as well as through the desquamation of cells from the gastrointestinal tract.

Of note is that in children and growing animals, there is necessarily a greater input of cholesterol into the body than output since there is a net accumulation of cholesterol to maintain an increased body weight. However, once adulthood is reached and body weight becomes constant, the input of cholesterol into the system must be equal to the output.

Overall, the regulatory mechanisms on cholesterol metabolism must be operative, which accurately adjust the rate of cholesterol biosynthesis within the body and the rate of cholesterol excretion from the body to accommodate the varying amounts of cholesterol that are absorbed by the intestine at different times. In general, these regulatory mechanisms on cholesterol metabolism function well, so there is little net accumulation of cholesterol in the body, and yet sufficient cholesterol is always available to meet the metabolic needs of the various tissues. However, in some species, and in particular humans, delicate imbalances result in an increased cholesterol concentration in the plasma and/or induce hepatic hypersecretion of cholesterol into bile [102–105]. In the cardiovascular system, this metabolic abnormality leads to an accumulation of the cholesteryl ester molecule in cells within the walls of arteries, which causes clinically apparent atherosclerosis [25, 35, 106–111]. In the biliary system, when an imbalance takes place, gallbladder bile become supersaturated with cholesterol, thus leading to the precipitation of solid plate-like cholesterol monohydrate crystals, and eventually, to clinically apparent cholesterol cholelithiasis [112–122].

CHOLESTEROL SYNTHESIS RATES

The daily dietary cholesterol that is absorbed through the small intestine provides the first major source for sterol in the body pool. The *de novo* cholesterol biosynthesis by the major organs such as the liver and intestine is the second major source for the cholesterol pool in the body. The cholesterol synthesis rates in humans and experimental animals have been carefully investigated by two methods [49–51, 89, 96, 123–133]: (i) sterol balance analysis and (ii) measurement of the incorporation of [^{3}H]water into sterols. By measuring sterol balance across the body, the amount of cholesterol excreted daily from the body in the feces as neutral (cholesterol and its bacterial degradation products) and acidic (bile acids) sterols is quantitated in the steady state. In theory, based on these data, together with the daily dietary cholesterol intake, the cholesterol synthesis rates can be determined after correcting the amounts of cholesterol that are lost from the skin, converted to steroid hormones, and degraded by intestinal bacteria. Because it is not easy to directly detect these parameters that reflect the loss of cholesterol from the skin, its conversion to steroid hormones, and its degradation by intestinal bacteria, the values obtained by sterol balance methods may be a little bit higher than the actual rate of cholesterol synthesis within the body. Nevertheless, this method has been extensively used for human and animal studies to calculate the rate of total cholesterol synthesis per day in these experimental subjects. The second method is used mainly in animal experiments, which can determine the rate of whole body cholesterol synthesis directly by measuring the rate at which an animal incorporates [^{3}H]water into sterols under *in vivo* conditions [49–51, 54, 89, 124, 134]. These two methods can produce comparable results. It has been found that the rates of cholesterol synthesis are approximately 30–34 mg/day/kg body weight in the squirrel monkey and 110–118 mg/day/kg body weight in rats, respectively, regardless of whether it is measured by the sterol balance technique or by the [^{3}H]water incorporation procedure. Using such methods, it has been found that the rate of whole body cholesterol synthesis is approximately 8–10 mg/day/kg body weight in humans. These results suggest that humans and larger animals such as monkeys usually synthesize much less cholesterol per unit weight than do small animals such as rats and mice. It should be pointed out that the absolute amount of cholesterol synthesized varies significantly from animal to animal and is even very different between animals of similar weight.

In addition, [^{3}H]water is utilized to investigate the quantitative importance of the liver to whole body cholesterol synthesis mainly in experimental animals, particularly under in *vivo* conditions [49–51, 54, 89, 124, 134]. To determine the amount of newly synthesized cholesterol, the weight of each organ must be taken into consideration. Although the synthesis of cholesterol occurs in virtually all cells, the capacity is greatest in the liver, intestine, adrenal cortex, and reproductive tissues such as ovaries, testes, and placenta [52, 53, 92]. As shown in Figure 4.4, in the rat, approximately 50% of the newly synthesized cholesterol is found in the liver, 15% in the small intestine, 12% in the skin, and 13% in the carcass (mainly muscle and bone). It is also revealed that these organs provide the majority of the newly synthesized cholesterol to the body in other animal

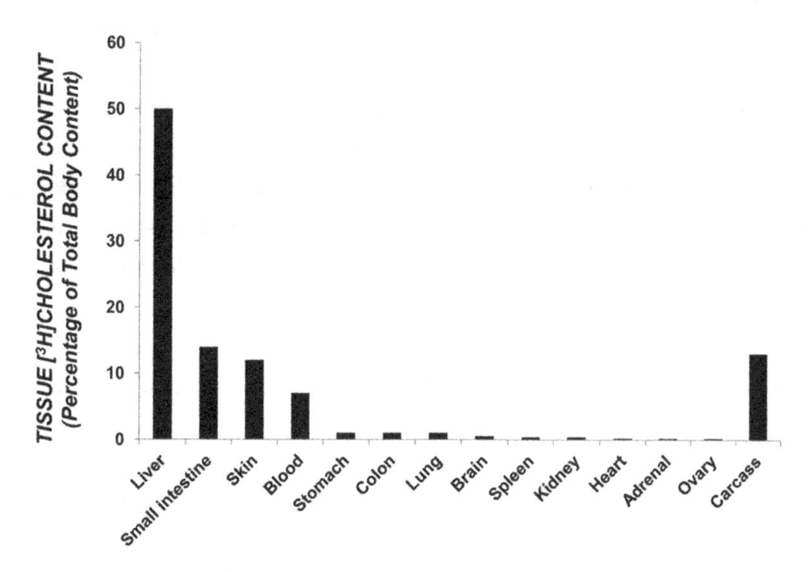

FIGURE 4.4: The importance of the major organ system as sites for cholesterol synthesis *in vivo* in the rat. The rat was administrated with [^{3}H]water intravenously and sacrificed 1 hour later. After the content of newly synthesized cholesterol in each major tissue is determined, it is expressed as a percentage of the rate of synthesis found in the whole animal.

species such as the monkey, hamster, and mouse. The quantitative importance of the liver to total body cholesterol synthesis has been investigated in different animal species under the condition of low dietary cholesterol intake. In the rat and squirrel monkey, approximately half of total body cholesterol synthesis comes from the liver. In the other species, however, hepatic cholesterol synthesis contributes less than one-third of the cholesterol synthesized in the body each day, approximately 30% in hamsters, and 20% in rabbits and guinea pigs. In humans, it has been established that the liver contributes approximately 10–15% of the newly synthesized cholesterol to the body. Of special note, when the amount of cholesterol in the diet is increased, these values would be proportionately reduced because cholesterol synthesis is suppressed. These changes are induced by a negative feedback regulation of cholesterol synthesis. However, it has been found that cholesterol feeding has little to no effect upon total body cholesterol synthesis rate in the human. It is likely that the liver of humans is not very sensitive to the feedback inhibition of cholesterol synthesis by dietary cholesterol or the liver makes only a small contribution to total body cholesterol synthesis.

CHOLESTEROL BIOSYNTHESIS PATHWAYS

Several early studies found that in rats fed acetate isotopically labeled in its carbon atoms, the newly synthesized cholesterol contained the isotopic label. This important experiment provided direct evidence showing that acetate is a precursor of cholesterol. In fact, during the cholesterol biosynthesis, all 27 carbons of the cholesterol molecule are derived from one precursor, acetyl CoA. By studying cholesterol that was synthesized from acetate labeled in either its methyl or its carboxyl carbon, it revealed that degradation of the isotope-labeled cholesterol indicated the origin of each atom of the cholesterol molecule. The second crucial discovery came from studies that mevalonate was a key step in the pathway of cholesterol synthesis because this C_6 acid could decarboxylate to produce the C_5 isoprene intermediate. Further studies found that the activated isoprene intermediate was isopentenyl pyrophosphate. Thus, a pathway for the synthesis of cholesterol from acetate can be outlined below:

$$\underset{C_2}{\text{Acetate}} \rightarrow \underset{C_6}{\text{Mevalonate}} \rightarrow \underset{C_5}{\text{Isopentenyl Pyrophosphate}} \rightarrow \underset{C_{30}}{\text{Squalene}} \rightarrow \underset{C_{27}}{\text{Cholesterol}}$$

The site for cholesterol biosynthesis is located mainly in the cytosol and endoplasmic reticulum. Of note, acetyl CoA can be obtained from several sources, including the β-oxidation of fatty acids, the oxidation of ketogenic amino acids such as leucine and lysine, and the pyruvate dehydrogenase reaction. These alterations suggest that cholesterol synthesis involves multimolecular interactions, as well as significant reducing power that is supplied in the form of NADPH. The latter is provided by glucose 6-phosphate dehydrogenase and 6-phosphogluconate dehydrogenase of the hexose monophosphate shunt pathway. The process of cholesterol synthesis requires hydrolysis of high-energy thioester bonds of acetyl CoA and phosphoanhydride bonds of ATP.

The Conversion of Three Acetyl CoA Molecules to Mevalonate

The first reaction in cholesterol synthesis is the formation of the intermediate mevalonate from acetyl CoA (Figure 4.5). Acetyl CoA is generated mainly from three sources: (i) β-oxidation of fatty acids, (ii) oxidation of ketogenic amino acids, and (iii) pyruvate dehydrogenase reaction. Also, acetate can be converted to acetyl CoA at the expense of ATP by acetokinase, or acetate thiokinase. The synthesis of mevalonate is the committed, rate-limiting step in cholesterol formation. In this cytoplasmic pathway, two molecules of acetyl CoA condense to yield acetoacetyl CoA by acetoacetyl-CoA thiolase (acetyl-CoA:acetyl-CoA acetyltransferase). Subsequently, acetoacetyl CoA is further condensed with a third molecule of acetyl CoA to form the 6-carbon compound 3-hydroxy-3-methylglutaryl CoA (HMG CoA) by HMG-CoA synthase (3-hydroxy-3-methylutarylyl CoA: acetoacetyl-CoA lyase) in the cytosol. However, this reaction is distinct from the mitochondrial HMG-CoA

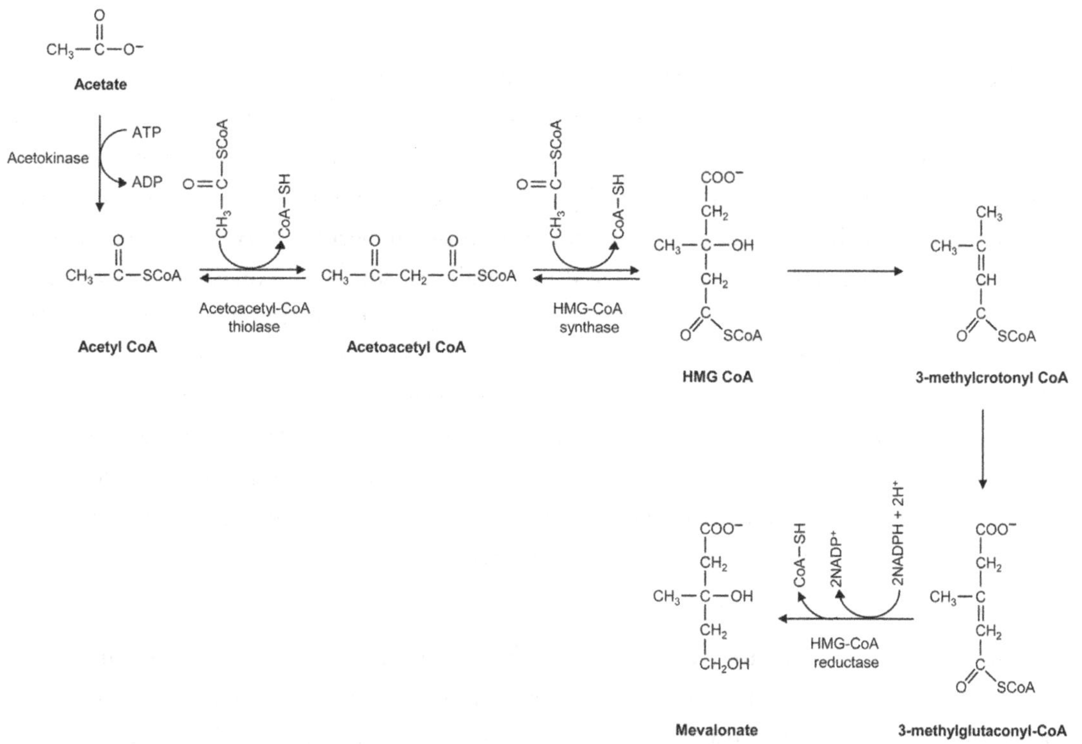

FIGURE 4.5: The conversion of three acetyl CoA molecules to mevalonate.

synthase, in where this enzyme catalyzes HMG CoA synthesis involved in production of ketone bodies. Also, HMG CoA is formed from oxidative degradation of the branched-chain amino acid leucine, through the intermediates 3-methylcrotonyl CoA and 3-methylglutaconyl CoA.

The formation of mevalonate involves the reduction of HMG CoA, a reaction that is catalyzed by the endoplasmic reticulum enzyme HMG-CoA reductase (mevalonate: $NADP^+$ oxidoreductase) [135]. HMG-CoA reductase contains eight membrane-spanning domains, and the amino-terminal domain, which faces the cytoplasm, contains the enzymatic activity. The reduction needs two molecules of NADPH for this process that induces hydrolysis of the thioester bond of HMG CoA. As a result, the primary alcohol group of mevalonate is formed. Because this reduction is irreversible and produces (R)-(+) mevalonate, HMG-CoA reductase catalyzes the rate-limiting reaction in cholesterol biosynthesis [84, 87, 88, 135–137]. HMG-CoA reductase is an intrinsic protein of the endoplasmic reticulum with its catalytic C-terminus domain extending into the cytosol. Phosphorylation of HMG-CoA reductase can reduce its catalytic activity and promote its degradation by increasing this enzyme sensitive to proteolytic action. Increased intracellular cholesterol can trigger the phosphorylation of HMG-CoA reductase so that the reduction of HMG CoA to mevalonate is reduced.

Synthesis of Two Activated Isoprenes from Mevalonate

The synthetic reactions that convert mevalonate to farnesyl pyrophosphate are involved in the transfer of three phosphate groups from three molecules of ATP to mevalonate (Figure 4.6). These transfers can activate both carbon 5 and the hydroxyl group on carbon 3 for further reactions. The stepwise transfer of the terminal phosphate group from two molecules of ATP to mevalonate to form 5-pyrophosphate mevalonate is catalyzed by mevalonate kinase and phosphomevalonate kinase. Decarboxylation of 5-pyrophosphate mevalonate by pyrophosphomevalonate decarboxylase produces a double bond in the 5-carbon product, Δ^3-isopentenyl pyrophosphate, the first of two activated isoprenes that are necessary for the synthesis of cholesterol. In this ATP-dependent reaction, ADP, P_i, and CO_2 are produced. It is thought that decarboxylation–dehydration proceeds by way of the intermediate 3-phosphomevalonate 5-pyrophosphate. Isopentenyl pyrophosphate is converted to its allylic isomer 3,3-dimethylallyl pyrophosphate by isopentenyl pyrophosphate isomerase in a reversible reaction, which is the second activated isoprene.

Synthesis of the 30-Carbon Squalene from Six Activated 5-Carbon Isoprenes

The head-to-tail condensation of 3,3-dimethylallyl pyrophosphate and Δ^3-isopentenyl pyrophosphate generates geranyl pyrophosphate (Figure 4.7). In this reaction, the "head" refers to the end of

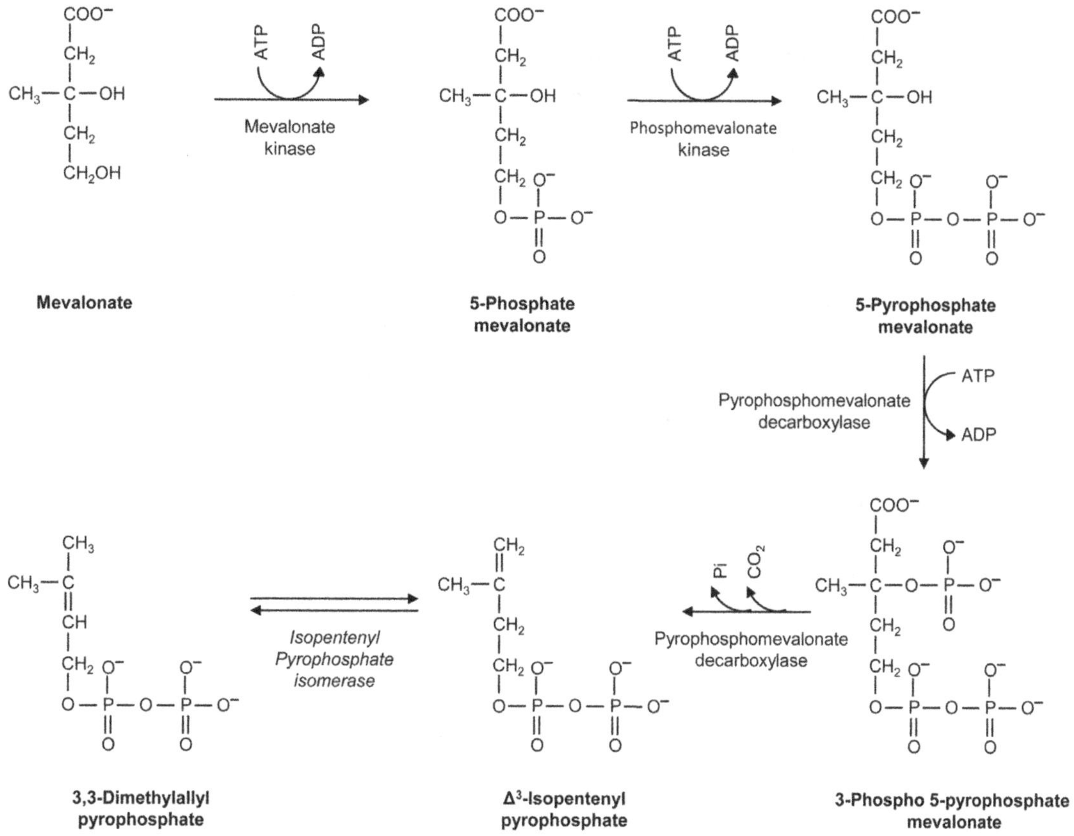

Mevalonate

Mevalonate kinase

5-Phosphate mevalonate

Phosphomevalonate kinase

5-Pyrophosphate mevalonate

Pyrophosphomevalonate decarboxylase

3-Phospho 5-pyrophosphate mevalonate

Pyrophosphomevalonate decarboxylase

Δ^3-Isopentenyl pyrophosphate

Isopentenyl Pyrophosphate isomerase

3,3-Dimethylallyl pyrophosphate

FIGURE 4.6: Synthesis of two activated isoprenes from mevalonate.

the molecule to which pyrophosphate is linked. After the pyrophosphate group of 3,3-dimethylallyl pyrophosphate is displaced, geranyl pyrophosphate, a compound with a 10-carbon chain, is formed. Geranyl pyrophosphate then undergoes another head-to-tail condensation with Δ^3-isopentenyl pyrophosphate, resulting in the formation of the 15-carbon intermediate, farnesyl pyrophosphate. The stepwise condensation of three C_5 isopentenyl units to form the C_{15} farnesyl pyrophosphate is catalyzed by geranyltransferase. Furthermore, the head-to-head fusion of two molecules of farnesyl pyrophosphate and the removal of both pyrophosphate groups generate squalene. This process is regulated by squalene synthase of endoplasmic reticulum, which releases two pyrophosphate groups and requires NADPH. Squalene contains 30 carbons with 24 in the main chain and 6 in the methyl group branches.

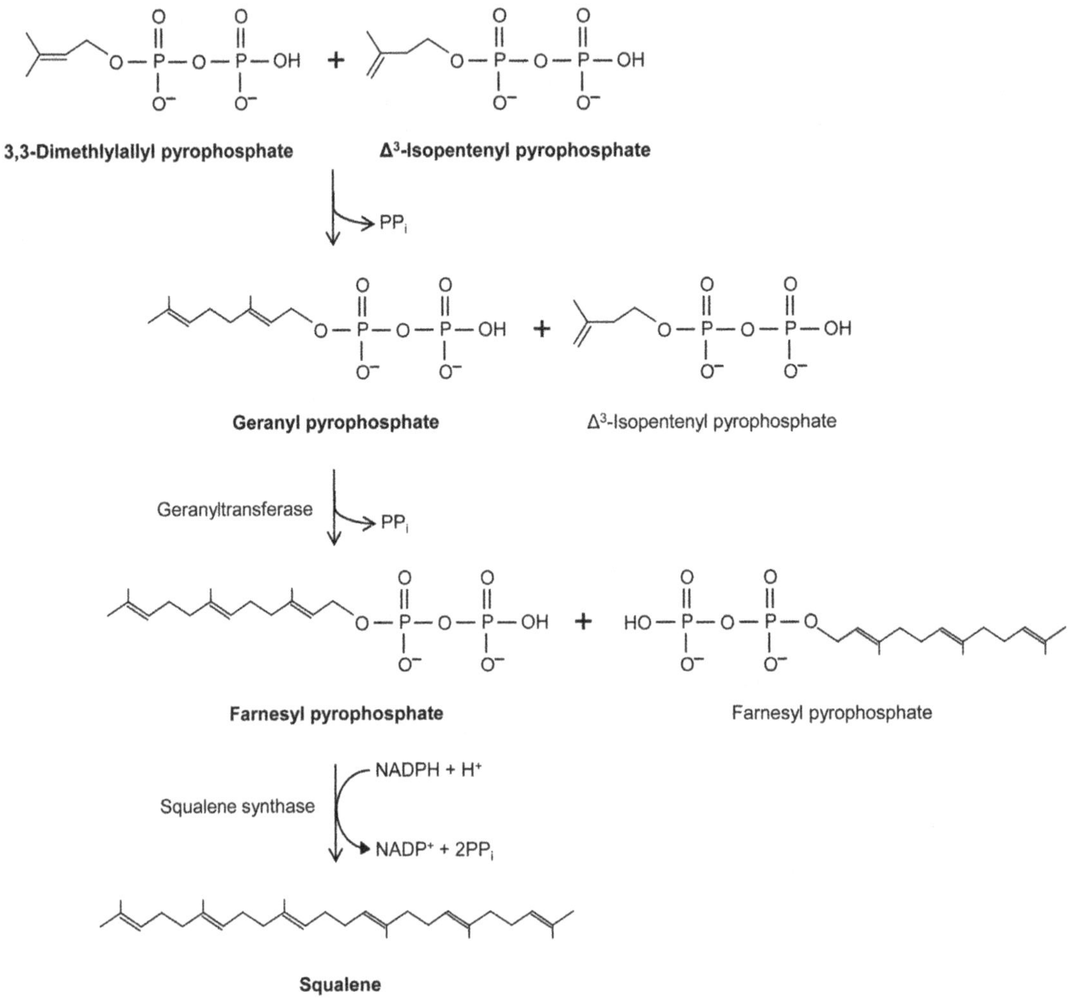

3,3-Dimethlylallyl pyrophosphate + **Δ^3-Isopentenyl pyrophosphate**

→ PP_i

Geranyl pyrophosphate + Δ^3-Isopentenyl pyrophosphate

Geranyltransferase → PP_i

Farnesyl pyrophosphate + Farnesyl pyrophosphate

Squalene synthase: NADPH + H^+ → $NADP^+$ + $2PP_i$

Squalene

FIGURE 4.7: Synthesis of the 30-carbon squalene from six activated 5-carbon isoprenes.

Cholesterol Is Synthesized from Squalene via Lanosterol

Cholesterol synthesis from squalene proceeds through the intermediate lanosterol, which contains the fused tetracyclic ring system and an eight-carbon side chain (Figure 4.8). The unsaturated carbons of the squalene 2,3-epoxide are aligned in a way that allows conversion of the linear squalene epoxide into a cyclic structure. The enzyme that catalyzes this reaction is bifunctional and contains squalene epoxidase or monooxygenase and a cyclase (lanosterol cyclase) activity. The hydroxylation of C-3 induces the cyclization of squalene to form lanosterol, a sterol with the four-ring structure characteristic of the steroid nucleus. Moreover, squalene monooxygenase adds a single oxygen atom from O_2 to the end of the squalene molecule, forming a squalene 2,3-epoxide. NADPH then

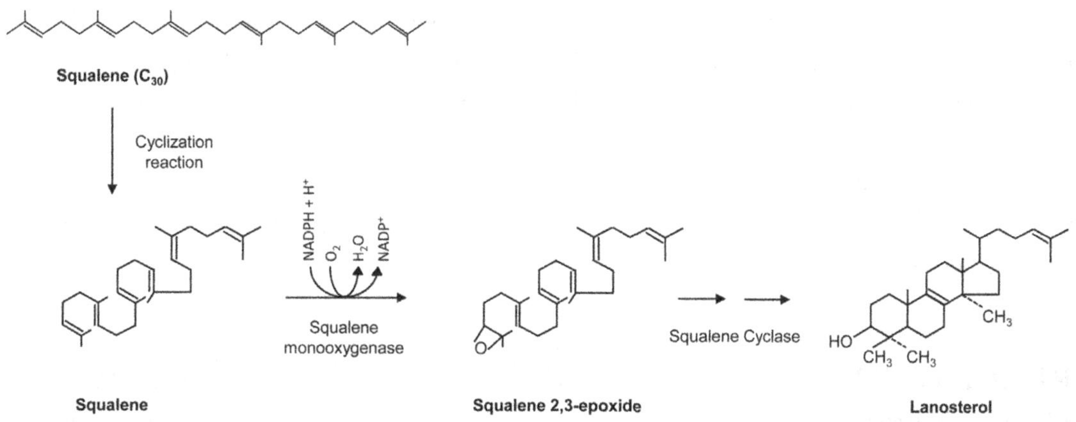

FIGURE 4.8: The conversion of squalene to lanosterol. The two molecules on the left show squalene in a different conformation, indicating how the cyclization reaction occurs.

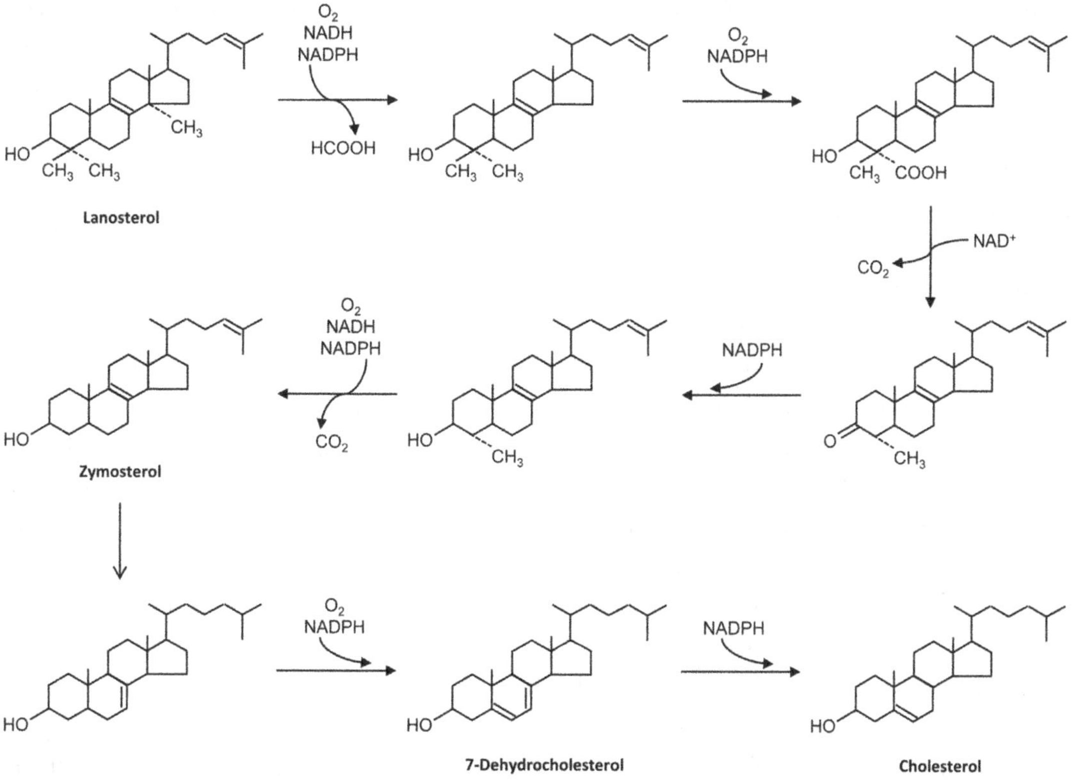

FIGURE 4.9: Cholesterol is synthesized from lanosterol. Used with permission from *Textbook of Biochemistry with Clinical Correlations*. Editor: Devlin TM. 6th edition. Wiley-Liss, Hoboken, NJ, 2006. p. 712.

reduces the other oxygen atom of O_2 to H_2O. Of special note is that transformation of lanosterol to cholesterol involves many steps and several enzymes (Figure 4.9), and these reactions include: (i) reduction of the double bond between the C-24 and C-25 positions in the side chain; (ii) deletion of two methyl groups at the C-4 position; (iii) deletion of the methyl group at the C-14 position; and (iv) relocation of the double bond between C-8 and C-9 to a position between C-5 and C-6. It should be noted that the OH group of lanosterol is in the β orientation, which projects above the plane of the A ring.

REGULATION OF HEPATIC CHOLESTEROL BIOSYNTHESIS

Although cholesterol is a crucial component of cellular membranes, it can be cytotoxic when excess amounts of cholesterol are accumulated within the cell. Therefore, cholesterol biosynthesis is precisely regulated by a negative feedback mechanism to maintain the appropriate cellular cholesterol level. Cholesterol synthesis is catalyzed by a large group of microsomal enzymes including acetoacetyl-CoA thiolase, HMG-CoA synthase, mevalonate kinase, phosphomevalonate kinase, farnesyl diphosphate synthase, squalene synthase, lanosterol 14α-demethylase, lathosterol oxidase, and 7-dehydrocholesterol reductase. Among these regulatory enzymes, it is well known that HMG-CoA reductase commits the rate-limiting step in cholesterol biosynthesis. These enzymes are regulated at the transcriptional level, and their transcriptional regulation is controlled by a family of transcription factors, sterol regulatory element-binding proteins (SREBPs). It has been found that SREBPs belong to a large class of transcription factors containing the basic helix–loop–helix–leucine zipper (bHLH-Zip) domains. Unlike other members of this class, SREBPs are synthesized as membrane-bound precursors that require a two-step proteolytic process of cleavage in order to release their amino-terminal bHLH-Zip-containing domains into the nucleus to bind to a specific DNA sequence, sterol regulatory element (SRE), and activate their target genes in a sterol-regulated manner. Some *in vitro* experiments and mouse studies have found that although there are three SREBP isoforms, designated SREBP-1a, -1c, and -2, SREBP-2 plays an active role in regulating the transcription of genes involved in cholesterol biosynthesis. SREBP-1c more specifically activates fatty acid synthesis. SREBP-1a is a potent activator of all SREBP-responsive genes, including those that mediate the synthesis of cholesterol, fatty acids, and triglycerides. Furthermore, SREBP-1a and SREBP-2 are the predominant isoforms of SREBP in most cultured cell lines, whereas SREBP-1c and SREBP-2 predominate in the liver and most other intact tissues. As shown in Figure 4.10, SREBP-2 regulates cholesterol synthesis very efficiently by controlling gene transcription of almost all of the genes for cholesterogenic enzymes in the cholesterol synthetic pathway. Indeed, some mouse experiments revealed that overexpression of the *SREBP-2* gene in the liver induces a significant increase in cholesterol biosynthesis.

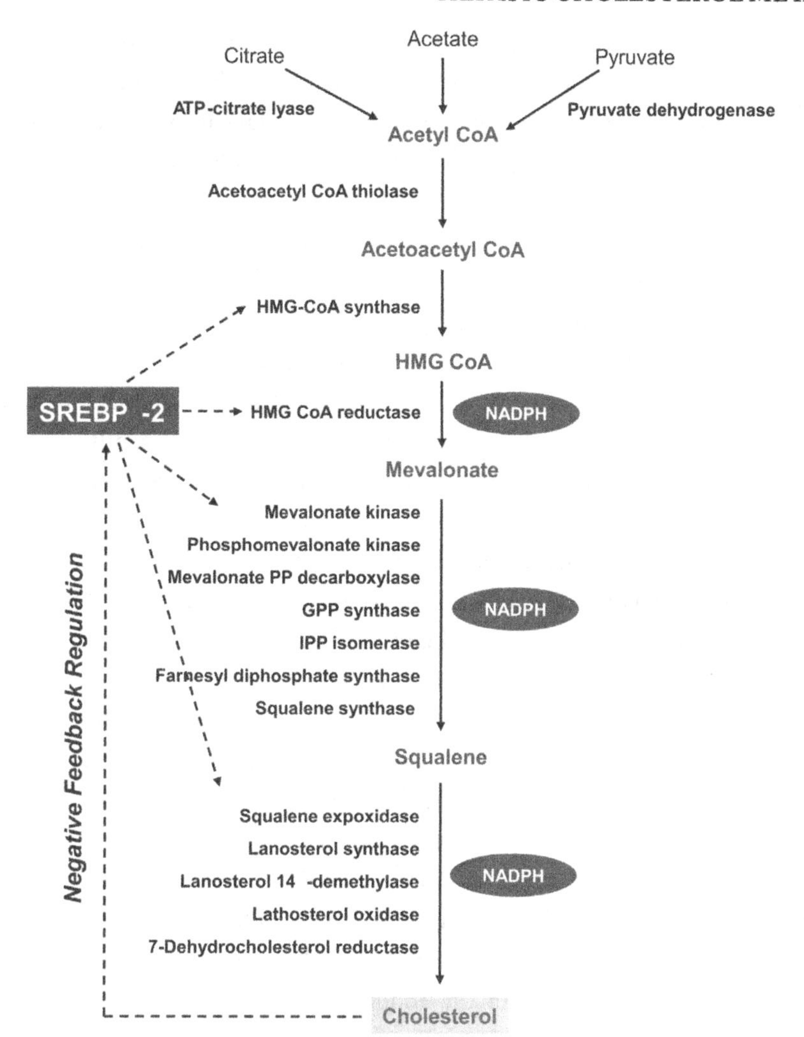

FIGURE 4.10: Genes regulated by sterol regulatory element-binding protein 2 (SREBP-2). The diagram shows the major metabolic intermediates in the pathways for synthesis of cholesterol. In the liver, SREBP-2 preferentially regulates cholesterol synthesis very efficiently by controlling gene transcription of the genes for cholesterogenic enzymes by a negative feedback mechanism. Of special note is that although is not indicated in this diagram, SREBP-2 can directly influence the expression of almost all of the genes in the cholesterol synthetic pathway. Abbreviations: GPP, geranylgeranyl pyrophosphate synthase; HMG CoA, 3-hydroxy-3-methylglutaryl coenzyme A; IPP, isopentenyl pyrophosphate; NADPH, nicotinamide adenine dinucleotide phosphate; PP, pyrophosphomevalonate.

As discussed in the section of cholesterol balance, the cholesterol pool of the body is derived from the intestinal sources of dietary and biliary cholesterol and cholesterol biosynthesis primarily in the liver and intestine. However, when dietary cholesterol intake is increased, cholesterol synthesis is markedly reduced [138–140]. In contrast, when dietary cholesterol consumption is reduced, cholesterol synthesis is significantly increased in the liver and intestine. These findings indicate that cholesterol exerts a negative feedback regulation on cholesterol synthesis, as well as induces feedback inhibition on HMG-CoA reductase and promotes the degradation of this enzyme. Because HMG-CoA reductase is the rate-limiting enzyme for catalyzing an irreversible step on the synthesis of mevalonate, it plays a central role in cholesterol biosynthesis [135]. It has been found that statins (e.g., lovastatin, pravastitin, fluvastatin, cerivastatin, atorvastatin, and simvastatin) can inhibit HMG-CoA reductase activity, particularly in the liver, and significantly reduce plasma total and LDL cholesterol by as much as 20–50% in patients with hypercholesterolemia or dyslipidemia [141].

Figure 4.11 displays a model of the SREBP pathway and elucidates how this pathway is determined by one of its major end-products, cholesterol at a molecular level [142]. Regulation of SREBPs occurs at two levels—transcriptional and posttranscriptional. The transcriptional regulation of the SREBP-2 is a complex process [136, 137, 142–145]. To reach the nucleus and act as a transcription factor, the NH_2-terminal domain of SREBP-2 must be released from the membrane proteolytically. Three proteins are required for this processing: an escort protein, SREBP cleavage-activating protein (SCAP), and two proteases, Site-1 protease (S1P) and Site-2 protease (S2P). Newly synthesized SREBP-2 is inserted into the membranes of the endoplasmic reticulum, where its COOH-terminal regulatory domain binds to the COOH-terminal domain of SCAP. Moreover, SCAP is both a sensor of sterols and an escort for SREBP-2. When cellular cholesterol is reduced, SCAP senses the shortage of cholesterol in the cell through its membranous sterol-sensing domain [146]. SCAP transports SREBP-2 from the endoplasmic reticulum to the Golgi apparatus. In the Golgi apparatus, S1P, a membrane-bound serine protease, cleaves SREBP-2 in the luminal loop between its two membrane-spanning segments, dividing the SREBP-2 molecule in half. The NH_2-terminal bHLH-Zip domain is then released from the membrane via a second cleavage mediated by S2P. The nuclear SREBP (nSREBP) translocates to the nucleus of cell, where it activates transcription by binding to sterol response elements (SREs) in the promoter/enhancer regions of multiple target genes for cholesterogenic enzymes in the cholesterol synthetic pathway. When the cholesterol concentration of cells rises, SCAP senses the excess cholesterol through its membranous sterol-sensing domain. As a result, SCAP changes its conformation and the SCAP/SREBP-2 complex is no longer incorporated into transport vesicles in the endoplasmic reticulum [147]. Under the circumstances, SREBP-2 is not able to access S1P and S2P in the Golgi apparatus, so their bHLH-Zip domains cannot be released from the membrane of endoplasmic reticulum. Therefore, the

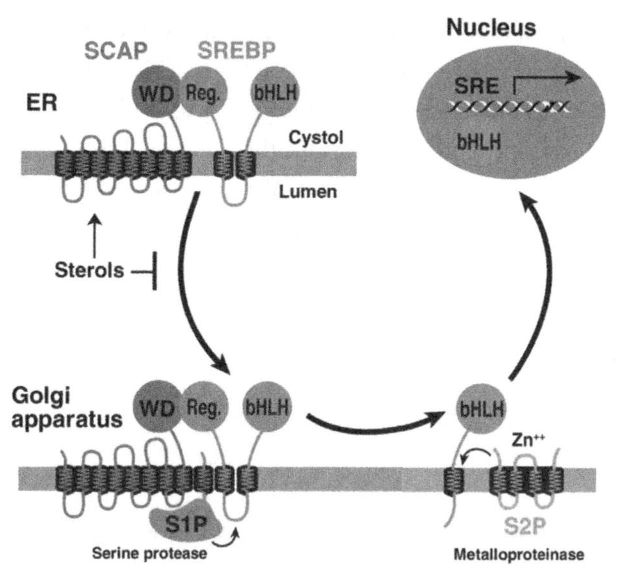

FIGURE 4.11: The activation of sterol regulatory element-binding proteins (SREBPs) through proteolytic processing for regulating cholesterol biosynthesis. The SREBP cleavage-activating protein (SCAP) is both a sensor of sterols and an escort of SREBP-2. When the cholesterol concentration of cells is reduced, SCAP senses the shortage of cholesterol in the cell through its membranous sterol-sensing domain. Then, SCAP moves SREBP-2 from the endoplasmic reticulum (ER) to the Golgi apparatus, where two proteases, Site-1 protease (S1P) and Site-2 protease (S2P), act sequentially to release the NH_2-terminal the basic helix–loop–helix–leucine zipper (bHLH-Zip) domain from the membrane. The bHLH-Zip domain enters the nucleus and binds to a sterol response element (SRE) in the enhancer/promoter region of target genes, thus activating their transcription. When cellular cholesterol content rises, SCAP senses the excess cholesterol. Consequently, the SCAP/SREBP-2 complex is no longer incorporated into ER transport vesicles. SREBP-2 is not able to reach the Golgi apparatus, and the bHLH-Zip domain cannot be released from the membrane. Under the circumstances, the transcription of all target genes for cholesterogenic enzymes in the cholesterol synthetic pathway is inhibited. Of special note is that the SREBP-2 can regulate not only the transcription of HMG CoA reductase, but also the transcription of the genes encoding many other enzymes in the cholesterol biosynthetic pathway, as shown in this figure, including HMG CoA synthase, farnesyl diphosphate synthase, squalene synthase, and so on. Used with permission from Goldstein, JL, Rawson, RB, Brown, MS. *Archives of Biochemistry and Biophysics.* 2002;397(2):139–148.

transcription of all target genes for cholesterogenic enzymes is inhibited. It should be emphasized that the SREBP-2 can regulate not only the transcription of HMG-CoA reductase, but also the transcription of the genes encoding many other enzymes in the cholesterol biosynthetic pathway, as shown in Figure 4.11, including HMG-CoA synthase, farnesyl diphosphate synthase, squalene synthase, and so on. Of note, the posttranscriptional regulation of SREBP-2 is a simple process, which involves the sterol-mediated suppression of SREBP-2 cleavage. These alterations result from the sterol-mediated suppression of the movement of the SCAP/SREBP-2 complex from the endoplasmic reticulum to the Golgi apparatus. Overall, cholesterol biosynthesis is tightly regulated by a negative feedback regulation mechanism through the SREBP-2 pathway in the liver.

• • • •

CHAPTER 5

Physical Chemistry and Hepatic Metabolism of Bile Acids

CHEMICAL STRUCTURE AND PHYSICAL–CHEMICAL PROPERTIES OF BILE ACIDS

In humans, the common bile acids possess a steroid nucleus of four fused hydrocarbon rings with polar hydroxyl functions and an aliphatic side chain conjugated in amide linkage with glycine or taurine [148–150]. The nucleus has 19 carbon atoms, 17 in the ring system and 2 in angular methyl groups (Figure 5.1). The only nuclear substituent occurring in most natural bile acids is the hydroxyl group that is at the 3, 6, 7, or 12 positions. Because natural dihydroxy bile acids, when ionized, are fully soluble at body temperature, and the acids of all trihydroxy bile acids are soluble, they are classified as soluble amphiphiles [46, 150–155]. The common bile acids differ in the number and orientation of the hydroxyl groups on the steroid nucleus [46, 154–158]. The structures of the common bile acids in humans are summarized in Table 5.1. Technically, the term bile acid refers to the form in which the carboxylic acid side chain is protonated (non-ionized), and bile salt refers to the ionized form. The two forms coexist in aqueous solution, but at a physiological pH of 7.4 the acid form predominates. In this chapter, the terms bile acid and bile salt are used interchangeably.

Natural bile acids have a hydrophilic face (α side) and a hydrophobic face (β side). The hydrophilic (polar) areas of bile acids are the hydroxyl groups and the conjugation side chain of either glycine or taurine, and their hydrophobic (nonpolar) area is the ringed steroid nucleus (Figure 5.2). Due to the possession of both hydrophilic and hydrophobic surfaces, bile acids are highly soluble, detergent-like amphiphilic molecules. Their high aqueous solubility is due to their capacity to self-assemble into micelles when a critical micellar concentration (CMC) is exceeded [151, 152, 159–165]. In natural bile acids, the hydrophobic side is large, and the CMC value (in 0.15 M Na^+) is less than 10 μM. As the contiguous area of the hydrophobic side is decreased, the CMC increases. The CMC of 5β bile acids increases when α substituents are changed to β substituents or to oxo groups. Under normal physiological conditions, the CMC values for common bile acids are between 1 and 20 μM, which is dependent on the species of bile acids, the ionic strength and composition, and the types and concentrations of other lipids present in solution. Because bile is concentrated

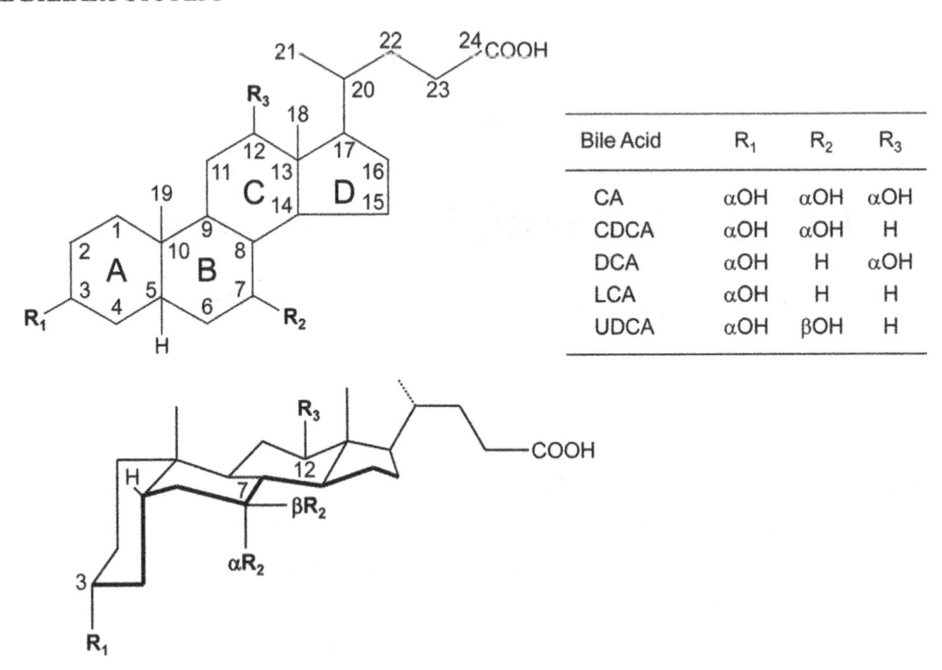

Bile Acid	R_1	R_2	R_3
CA	αOH	αOH	αOH
CDCA	αOH	αOH	H
DCA	αOH	H	αOH
LCA	αOH	H	H
UDCA	αOH	βOH	H

FIGURE 5.1: Molecular structure of common bile acids, showing common steroid ring and side-chain structure. The numbering of the carbon atoms in bile acids is shown in the structural formula (top panel). Hydroxyl group(s) location and orientation are given for each bile acid (bottom panel). Abbreviations: CA, cholic acid; CDCA, chenodeoxycholic acid; DCA, deoxycholic acid; LCA, lithocholic acid; UDCA, ursodeoxycholic acid.

gradually within the biliary tree, bile acid concentrations steadily exceed their CMCs so that bile acids can form simple micelles in bile [26, 28, 46, 154, 162, 166–173]. Of note, micelles of bile acids can solubilize other types of lipids such as cholesterol and phospholipids by forming mixed micelles in bile. The lower the CMC is, the better the ability of the bile acids is to solubilize polar lipids.

Bile acids are amidated with glycine or taurine [174–178]. It would be expected that the elongation of the side chain with glycine or taurine would lower the CMC. In fact, amidation with glycine or taurine has little or no effect on the CMC because the additional lipophilicity of the ethylene groups in glycine or taurine appears to be balanced by the additional hydrophilicity of the amide bond. An explanation for the specificity of amino acid conjugation of bile acids, i.e., with glycine or taurine and not with other amino acids, has been proposed. Conjugates of cholic acid with amino acids other than glycine or taurine have been shown to be cleaved rapidly by pancreatic carboxypeptidases. Thus, if bile acids are amidated with amino acids other than glycine or taurine, they would be hydrolyzed by pancreatic enzymes during small intestinal transit.

TABLE 5.1: Biliary bile salt composition in healthy humans.

BILE SALT		PERCENT IN BILE [a]		PERCENT OF TOTAL	
TRIVIAL NAME	SYSTEMATIC NAME	TAURINE CONJUGATES	GLYCINE CONJUGATES	SULFATED	GLUCURONIDATED
Common [b]					
Cholate	3α,7α,12α-Trihydroxy-5β-cholanoate	12	23	0.1	0.1
Chenodeoxycholate	3α,7α-Dihydroxy-5β-cholanoate	12	23	0.2	0.1
Deoxycholate	3α,12α-Dihydroxy-5β -cholanoate	8	16	N.D. [c]	N.D.
Ursodeoxycholate	3α,7β-Dihydroxy-5β -cholanoate	trace - ~2	trace - ~4	N.D.	N.D.
Lithocholate	3α-Monohydroxy-5β-cholanoate	1 - ~2	3 - ~4	60–80	trace
Uncommon					
Isochenodeoxycholate	3β,7α-Dihydroxy-5β-cholanoate	trace - ~1.0		N.D.	N.D.
Isodeoxycholate	3p, 12α-Dihydroxy-5β-cholanoate	trace - ~0.3		N.D.	N.D.

TABLE 5.1: (*continued*)

BILE SALT		PERCENT IN BILE [a]		PERCENT OF TOTAL	
TRIVIAL NAME	SYSTEMATIC NAME	TAURINE CONJUGATES	GLYCINE CONJUGATES	SULFATED	GLUCURONIDATED
7-Oxo-chenodeoxycholate	3α-Monohydroxy-, 7-oxo-5β-cholanoate	trace - ~0.6		N.D.	N.D.
12-Oxo-deoxycholate	3α-Monohydroxy-, 7-oxo-5β-cholanoate	trace - ~1.7		N.D.	N.D.
7-Oxo-cholate	3α,12α-Dihydroxy, 7-oxo-5β-cholanoate	trace - ~0.8		N.D.	N.D.
12-Oxo-cholate	3α,7α-Dihydroxy, 12-oxo-5β-cholanoate	trace - ~0.8		N.D.	N.D.
3-Oxo-deoxycholate	12α-Monohydroxy, 3-oxo-5β -cholanoate	trace - ~0.05		N.D.	N.D.
3-Oxo-chenodeoxycholate	7α-Monohydroxy, 3-oxo-5β-cholanoate	trace - ~0.1		N.D.	N.D.
Ursocholate	3α,7 p ,12α-Trihydroxy-5β-cholanoate	Infrequent		N.D.	N.D.
	3β-Monohydroxy-Δ-5-cholanoate	trace		N.D.	N.D.

[a] Unconjugated bile salts comprise 0.1 to 0.4% of total biliary bile salts.
[b] Comprise 92 to 99% of total biliary bile salts.
[c] N.D. = not determined.
Used with permission from Carey MC and Cahalane MJ. Enterohepatic Circulation in *The Liver: Biology and Pathobiology*. Editors: Arias IM, Jakoby WB, Popper, Schachter D, and Shafritz DA. 2nd edition. Raven Press, New York, 1988. p. 575.

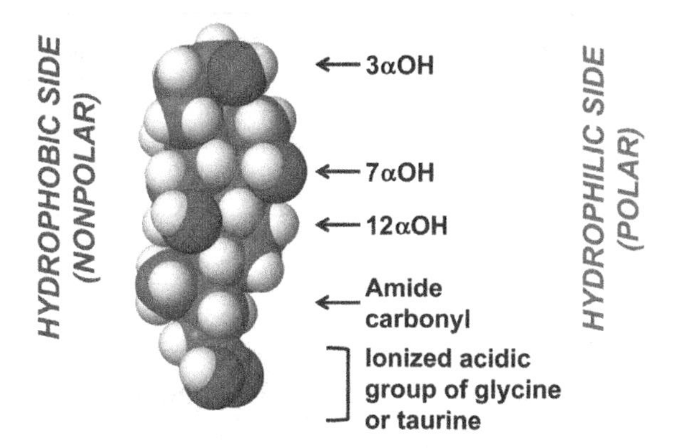

FIGURE 5.2: Space filling model of the taurine conjugate of cholic acid (viewed from the side) showing its planar amphipathic structure with a hydrophilic and a hydrophobic side.

Amidation has an important effect on the solution properties of bile acids [179–181]. The glycine conjugates are soluble down to pH 4 to 5, whereas unconjugated bile acids are insoluble below pH 6 to 7. Conjugated bile acids are probably far more resistant to precipitation by heavy metals (e.g., copper or iron) or alkaline earth metals (e.g., calcium). Thus, amidation results in the formation of a bile acid that remains fully in solution in the small intestinal lumen.

The biological significance of amidation of bile acids is mostly extrahepatic. Amidation prevents passive absorption of bile acids in the biliary tract and the small intestine and keeps bile acids in solution. These two effects promote a high intraluminal concentration of bile acids, which is essential for facilitating the digestion and absorption of intestinal fat. However, in some birds, all biliary bile acids are unconjugated. Thus, amidation is not essential for hepatic secretion of biliary bile acids. Nonetheless, amidation of the side chain does prevent further conjugation of the side chain during hepatic transport.

Table 5.2 summarizes the physiological and biochemical functions of bile acids in the liver, biliary tract, and intestine. The major functions of bile acids in the liver and gastrointestinal tract include: (i) to induce bile formation, bile acids promote hepatic secretion of biliary lipids such as cholesterol and phospholipids and solubilize these lipids in mixed micelles. The vectorial movement of bile acids from blood into the bile canaliculus generates an osmotic water flow and is a major determinant of bile formation. This process can regulate the hepatic secretion of the major bile components such as cholesterol, phospholipids, and bilirubin. Bile acids promote bile flow by their osmotic effects, which enhance bile acid-dependent bile flow. (ii) Bile acids play an obligatory role in intestinal absorption of cholesterol, fatty acids, and fat-soluble vitamins and play an important role in aiding the digestion of dietary fat. Bile acids promote intestinal absorption by solubilizing

TABLE 5.2: Function of bile acids in humans
Whole organism
Elimination of cholesterol
Liver
Hepatocyte
Insertion of canalicular bile acid and phospholipid transporters
Induction of bile flow and biliary lipid secretion
Promotion of mitosis during hepatic regeneration
Regulation of gene expression by activation of FXR
Stimulation of synthesis and secretion of FGF-15
Endothelial cells
Regulation of hepatic blood flow via activation of TGR5
Biliary tract
Lumen
Solubilization and transport of cholesterol and organic anions
Solubilization and transport of heavy metal cations
Cholangiocytes
Stimulation of bicarbonate secretion via CFTR and AE2
Promotion of proliferation when obstruction to bile flow
Gallbladder epithelium
Modulation of cAMP-mediated secretion
Promotion of mucin secretion

TABLE 5.2: (*continued*)
Small intestine
Lumen
Micellar solubilization of dietary lipids
Cofactor for bile salt dependent lipase
Antimicrobial effects
Enhancement of tryptic hydrolysis of dietary proteins
Ileal enterocyte
Regulation of gene expression via nuclear receptors
Release of FGF-15
Ileal epithelium
Secretion of antimicrobial factors (FXR- mediated)
Large intestine
Colonic epithelium and muscular layer
Promotion of defecation by increasing propulsive motility
Colonic enterocyte
Modulation of fluid and electrolyte absorption
Brown adipose tissue
Promotion of thermogenesis via TGR5

Abbreviations: AE2, chloride–bicarbonate anion exchanger isoform 2; CFTR, cystic fibrosis transmembrane regulator; FGF-15, fibroblast growth factor 15; FXR, farnesoid X receptor; TGR5, G protein-coupled receptor 5.
Used with permission from Hofmann AF and Hagey LR. *Cellular and Molecular Life Sciences*. 2008; 65(16): 2461–2483.

dietary lipids and their digestion products as mixed micelles to facilitate their aqueous diffusion across an unstirred water layer, a surface mucous coat, and the intestinal mucosa. (iii) Bile acids play a complex role in maintaining cholesterol homeostasis. On one hand, bile acids are required for facilitating intestinal absorption of dietary and biliary cholesterol and thus increase the delivery of intestinal cholesterol to the liver. On the other hand, bile acids function through several mechanisms to promote the elimination of cholesterol from the body. Bile acids are water-soluble products of cholesterol catabolism and contribute to the elimination of cholesterol via fecal excretion. They also facilitate hepatic secretion of cholesterol into bile by solubilizing biliary cholesterol, thereby enabling cholesterol to move from the liver to the intestine for elimination. (iv) Bile acids act as signaling molecules for regulating fatty acid, glucose, and energy metabolism. (v) Conjugated bile acids have antimicrobial effects in the lumen, and could also stimulate the ileal enterocyte to secrete undefined antimicrobial agents. (vi) Bile acids bind calcium and prevent the formation of calcium gallstones and oxalate kidney stones. (vii) Bile acids stimulate the release of fibroblast growth factor 19 (FGF19) from the ileal enterocyte, which influences gallbladder refilling postprandially.

HYDROPHILIC–HYDROPHOBIC BALANCE OF BILE ACIDS

The potency of bile acids as detergents depends critically upon the distribution and orientation of hydroxyl groups around the steroid nucleus of the molecule, which is usually conferred on its hydrophobicity [182–184]. Because bile acids have a range of polarity, a hydrophobic index value (HI) can be derived from relative retention time during high pressure liquid chromatography (HPLC) with a C_{18} octadecylsilane stationary phase [48, 184–187]. The hydrophobic index can be used to predict the biological effects of individual bile acids. Furthermore, because many differences in the biologic properties of bile acids result from their relative affinity for aqueous versus lipid environments, the hydrophilic–hydrophobic balance could be evaluated by determining the partition coefficient for bile acids between polar and nonpolar solvents or by measuring the retention of bile acids in a liquid chromatographic system employing a polar mobile phase and a nonpolar stationary phase. The physical–chemical properties of bile acids also depend upon the nature and ionization state of functional groups on the side chain. It has been found that the glycine conjugate is more hydrophobic than the taurine conjugate. However, the conjugation of the side chain with glycine or taurine has little influence on the hydrophobic activity of fully ionized bile acids. Moreover, the relative hydrophobic activity declines as the number of hydroxyl groups increases. However, the orientation of the hydroxyl groups is also important; hydroxyls that project in the plane of the sterol nucleus or toward its β surface tend to disrupt the contiguous hydrophobic surface and strongly reduce hydrophobic activity. Thus, as found by reverse-phase liquid chromatographic methods, the dihydroxy bile salts ursodeoxycholate (3α, 7β) and hyodeoxycholate (3α, 6α) are more hydrophilic than the trihydroxy bile salt cholate (3α, 7α, 12α). Non-ionized bile acids are more hydrophobic than are the corresponding ionized bile acids.

As a rule, the more hydrophobic a bile acid is, the greater its tendency to self-associate in aqueous media [188]. Consequently, more hydrophobic bile acids display lower CMCs and tend to form larger micelles (higher mean aggregation number). More hydrophobic bile acids also exhibit greater solubilizing capacity for cholesterol and phospholipids, and thus, they are better detergents [36, 104, 168, 171, 189, 190]. They are also more injurious to the membranes of cells, both *in vivo* and *in vitro*.

Natural bile acid pools invariably contain multiple types of bile acids [191–194]. Mixtures of two or more types of bile acids of differing hydrophobic activity may behave as a single bile acid of intermediate hydrophobic activity with regard to self-association properties. For example, equimolar mixtures of taurocholic acid and taurodeoxycholic acid exhibit CMC values and micellar aggregation numbers between those of the two individual bile acids. In addition, the detergent properties and the toxicity of mixtures of two bile acids often are intermediate between the individual components. This may account for the observation that ursodeoxycholic acid (UDCA) may ameliorate the cholestasis caused by a rapid high-dose infusion of more hydrophobic bile acids in rats [195–201]. It may also account for the proposed therapeutic benefit of UDCA in the treatment of chronic liver diseases [202–210]. A hydrophobic activity index quantifying the overall hydrophobic–hydrophilic balance of mixed bile acid solutions has been proposed and can be quantified by HPLC techniques [186].

BILE ACID BIOSYNTHESIS PATHWAYS

Bile acids are derived from cholesterol as shown in Figure 5.3, and bile acid biosynthesis is the predominant metabolic pathway for catabolism of cholesterol in humans. The liver is the only organ that has all the enzymes required for *de novo* biosynthesis of two primary bile acids, cholic acid (CA, 3α, 7α, 12α-trihydroxy-5β-cholanoate), a trihydroxy bile acid with hydroxyl groups at the C-3, C-7, and C-12 positions, and chenodeoxycholic acid (CDCA, 3α, 7α-dihydroxy-5β-cholanoate), a dihydroxy bile acid with hydroxyl groups at the C-3 and C-7 positions [184, 211–228]. Because the enterohepatic circulation of bile acids is highly efficient, it allows less than 0.5% (about 0.2–0.6 g per day) of the secreted bile acids to be lost in the feces [178, 229–232]. Therefore, most of the bile acids that are transported by the hepatocyte are "old" bile acids that were previously synthesized and secreted into bile. The half-life of bile acids in humans is 2 to 3 days, so the average bile acid molecule makes 10 to 20 cycles in the enterohepatic circulation before it is lost from the exchangeable bile acid pool. Thus, the mass of bile acids in the enterohepatic circulation is maintained by continuous bile acid biosynthesis from cholesterol.

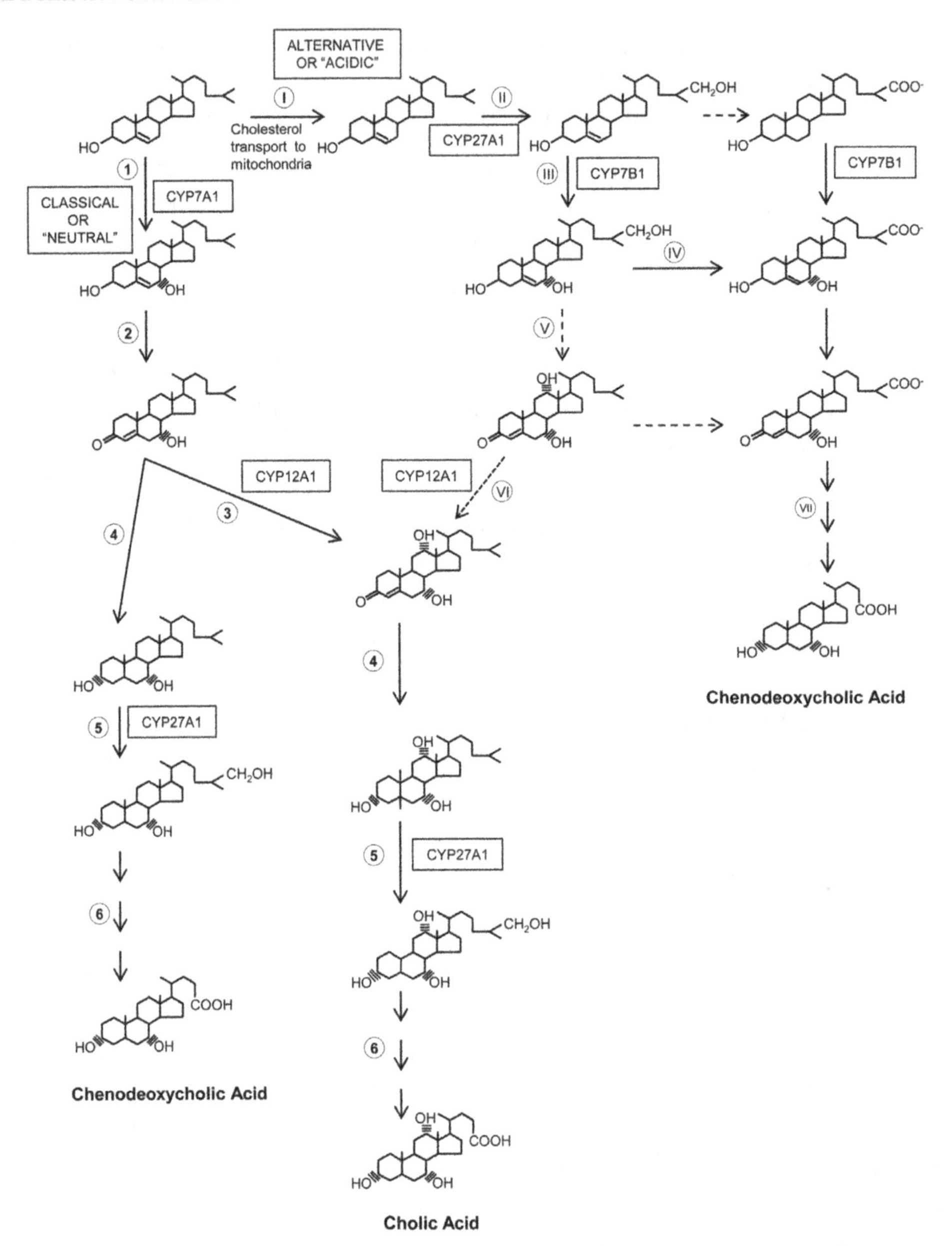

FIGURE 5.3: Bile acid biosynthetic pathways. Microsomal cholesterol 7α-hydroxylase (CYP7A1) (*step 1*) and mitochondrial sterol 27-hydroxylase (CYP27A1) (step II) catalyze initial reactions in the classic ('neutral') and alternative ('acidic') pathways of bile acid biosynthesis, respectively. **Classic pathway:** *step 1*, CYP7A1; *step 2*, 3β-hydroxysteroid oxidoreductase-isomerase; *step 3*, sterol 12α-hydroxylase (CYP12A1); *step 4*, Δ^4-3-oxo-steroid-5β-reductase and 3α-hydroxysteroid oxidoreductase; *step 5*,

The synthesis of bile acids from cholesterol in the liver involves a number of complex steps in modifying both the nucleus and the side chain. It involves at least 17 enzymes that make 25 possible intermediates. It is believed that these processes take place mainly in the pericentral hepatocyte of the hepatic acini. The bile acid synthesis from cholesterol is a superb example of the capacity of the hepatocyte to convert an insoluble lipid into a water-soluble and amphipathic compound. During bile acid synthesis, both the steroid nucleus and the side chain of cholesterol are greatly modified, inducing a remarkable change in the physical–chemical properties of cholesterol. Figure 5.3 illustrates the changes in chemical structure in the biotransformation of cholesterol, a lipophilic compound, to water-soluble products—the primary bile acids, CA and CDCA. Table 5.3 lists major enzymes for CA biosynthesis in the liver.

Hepatic bile acid biosynthesis involves two major pathways: the "classic" neutral pathway (i.e., cholesterol 7α-hydroxylase pathway) and the "alternative" acidic pathway (i.e., oxysterol 7α-hydroxylase pathway). In the classic neutral pathway, the rate-limiting enzyme, cholesterol 7α-hydroxylase (CYP7A1), converts cholesterol directly into 7α-hydroxycholesterol. In the alternative acidic pathway, cholesterol is transported to mitochondria where it must first be converted by C-24, C-25, or C-27 sterol hydroxylases into oxysterols, the major species being 27-hydroxycholesterol, before oxysterol 7α-hydroxylation [233–238]. Of the two major biosynthetic pathways, the neutral pathway is believed to be quantitatively more important in humans. This conclusion is supported by the finding that bile acid production has been reported to be decreased by almost 90% in an adult patient with an inherited CYP7A1 defect in the neutral pathway. In contrast, the acidic pathway may be dominant in neonates, as evidenced by the apparent lack of CYP7A1 in newborns and the finding of severe cholesteatic liver disease in an infant with an inherited oxysterol 7α-hydroxylase gene (CYP7B1) defect [239–241]. Other minor pathways initiated by 25-hydroxylase in the liver and 24-hydroxylase in the brain also may contribute to bile acid synthesis [234, 242, 243]. A non-specific 7α-hydroxylase (CYP7B1) expressed in all tissues is involved in the generation of oxidized metabolites (oxysterols), which may be transported to the liver and converted to CDCA.

CYP27A1; *step 6*, side-chain oxidation steps to form cholic acid (CA). **Alternative pathway**: *step II*, CYP27A1; *step III*, oxysterol 7α-hydroxylase (CYP7B1); *step IV*, conversion of 27-hydroxycholesterol to corresponding acid; *step V*, 3β-hydroxysteroid oxidoreductase-isomerase; *step VI*, CYP12A1; *step VII*, formation of chenodeoxycholic acid (CDCA). Solid arrows represent known and preferred intermediate steps in the two pathways of bile acid synthesis. The broken arrow represents a putative enzymatic step that has yet to be confirmed. Used with permission from Pandak WM, Hylemon PB, Bjorkhem I, Eggertsen G, Redford K, and Vlahcevic ZR. Regulation by Bile Acids of Hepatic Sterol 12α-Hydroxylase (CYP12A1) and Cholic Acid Synthesis in the Rat in *Bile Acids and Cholestasis*. Editors: Paumgartner G, Stiehl A, Gerok W, Keppler D, and Leuschner U. Kluwer Academic Publishers, Dordrecht. 1999. p. 60.

TABLE 5.3: Enzymes involved in cholic acid synthesis in humans.

SUBSTRATE	ENZYME	LOCATION
Cholesterol ↓	Cholesterol 7α- hydroxylase	Microsomes
7α-Hydroxycholesterol ↓	3β-hydioxy-5Δ-C27-steroid oxidoreductase	Microsomes
7α-hydroxy-4-cholesten-3-one ↓	12α- Hydroxylase	Microsomes
7α-12α-dihydroxy-4-cholesten-3-one ↓	4-3-oxosteroid-5β-redxictase and 3α-hydioxysteroid dehydrogenase	Soluble Soluble
5β-cholestane.3α, 7α,12α triol ↓	26-hydroxylase	Mitochondria (primarily)
5-cholestane. 3α,7α,l2α,26-tetrol ↓	Alcohol dehydrogenase andaldehyde dehydrogenase	Soluble Soluble
3α,7α,l2α.trihydroxy, 5β-cholestanoic acid ↓	Ligase, desaturase, and hydratase	Peroxysomes
3α,7α,l2α-trihydroxy- 5β-cholanic acid (cholic acid)		

Although the overall process of bile acid biosynthesis is complex, the enzymes for this process can be divided into two broad groups: one for performing modifications to the sterol ring structure (i.e., nuclear biotransformations), and the other for modifying the sterol side chain (i.e., side chain biotransformations). In the classic neutral pathway, sterol ring modifications precede side chain changes, but in the alternative acidic pathway, the side chain modifications occur before or during changes to the sterol ring structure (Figure 5.3). During nuclear biotransformations, cholesterol 7-hydroxylation is converted to form a 7α-hydroxy compound via an isomerase and a reductase to a key intermediate chole-7α-hydroxy-Δ4-3-one [233, 244–246]. This unsaturated oxo derivative is a substrate for the 12-hydroxylase, and 12-hydroxylation commits bile acid irreversibly to CA synthesis. Side chain biotransformations are believed to occur in the mitochondrion with the beginning of 26-hydroxylation, which is the first step of side chain oxidation. After 26-hydroxylation, there is oxidation to form the C-26 carboxylic acid and then β-oxidation in the conventional manner. This step involves the introduction of a double bond at C-24 followed by 24-hydroxylation and oxidative cleavage of a 3-carbon fragment (mediated by peroxisomal enzymes) to yield "mature" C-24 bile acids.

The traditional paradigm was that nuclear biotransformation preceded side chain biotransformation. If the side chain were biotransformed without any nuclear changes, the result would be a 3β-hydroxy-Δ-5-cholenoic acid. Indeed, this bile acid is a major constituent of amniotic fluid in humans, suggesting that during fetal life, the enzymes for side chain oxidation may mature more avidly than the enzymes for nuclear biotransformations. The occurrence of this bile acid proves that side chain oxidation can proceed normally despite any biotransformation of the cholesterol nucleus.

Before hepatic secretion into the bile canaliculus, both CA and CDCA are conjugated via their carboxyl group to the amino group of glycine or taurine. The conjugation of bile acids, i.e., N-acyl amidation with glycine or taurine, can reduce their toxicity and increase their solubility for secretion into bile. Also, conjugation enhances the hydrophilicity of these bile acids and the acidic strength of the side chain from a weak acid ($pK_a = 5.0$) to a strong acid ($pK_a = 3.9$ for the glycine conjugate and $pK_a = 2$ for the taurine conjugate). The conjugation to glycine or taurine could reduce the passive diffusion of bile acids across cell membranes during their transit through the biliary tree and small intestine before conjugated bile acids are absorbed mainly in the ileum where a specific bile acid transporter, apical sodium-dependent bile acid transporter (ASBT) is present [247–249]. It has been found that inherited defects in bile acid conjugation can result in fat-soluble vitamin malabsorption and steatorrhea [250]. These studies indicate that conjugation could greatly help maintain high intraluminal concentrations of bile acids down the length of the small intestine to facilitate the digestion and absorption of dietary fat, fat-soluble vitamins, and cholesterol.

Although most of the conjugated bile acids can be efficiently absorbed intact mainly in the ileum, a small fraction (approximately 15%) of bile acids is deconjugated by the endogenous bacterial flora during their passage down the distal small intestine. When bile acids escape intestinal absorption and pass into the colon, glyco- and tauro-conjugated CA and CDCA are deconjugated, and 7α-dehydroxylase activity in bacterial flora removes a 7α-hydroxy group to form secondary bile acids—deoxycholic acid (DCA, 3α, 12α-dihydroxy-5β-cholanoate), a dihydroxy bile acid with hydroxyl groups at the C-3 and C-12 positions and lithocholic acid (LCA, 3α-monohydroxy-5β-cholanoate), a monohydroxy bile acid with a hydroxyl group at the C-3 position. Enzymes and intermediates of bile acid transformation by intestinal microflora are shown in Figure 5.4. As a result, dehydroxylation of the primary bile acids CA and CDCA can reduce their aqueous solubility. DCA is nearly insoluble in water at the pH of the cecum, and LCA is virtually insoluble in water at body

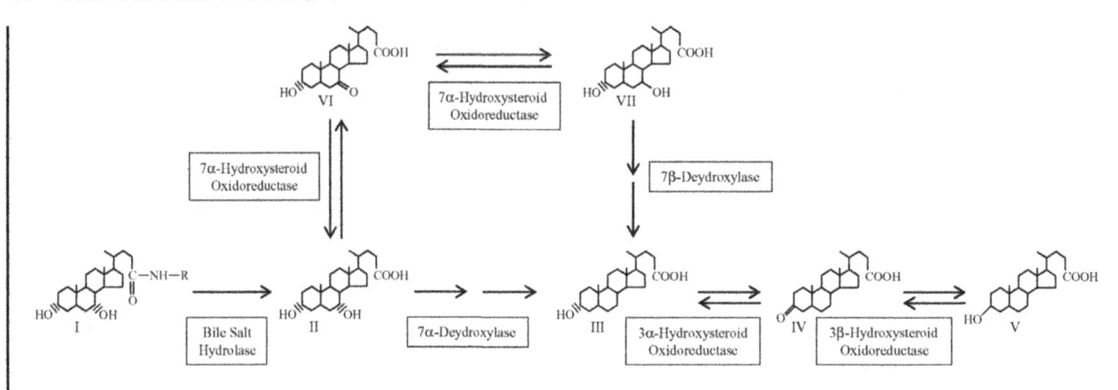

FIGURE 5.4: Enzyme and pathways of bile acid biotransformation by intestinal microflora. I = N-(3α, 7α-dihydroxy-5β-cholan-24-oyl) amino acid; II = chenodeoxycholic acid; III = lithocholic acid; IV = 3-oxo-5β-cholanoic acid; V = isolithocholic acid; VI = 3α-hydroxy-7-oxo-5β-cholanoic acid; and VII = ursodeoxycholic acid.

temperature. The colon absorbs approximately 50% of DCA and a small portion of LCA. After returning to the liver, DCA is reconjugated with glycine or taurine and circulates with the primary bile acids CA and CDCA. Consequently, the bile acid pool always contains DCA. Because hepatic conjugation of bile acids is very efficient, all of the circulating bile acids such as CA, CDCA, and DCA are in conjugated form. In contrast, because bacterial deconjugation and dehydroxylation in the colon are very efficient, all bile acids in the feces are in unconjugated form and are composed primarily of the secondary bile acids DCA and LCA.

In addition, the unconjugated bile acids are absorbed either actively or passively and returned to the liver, where they are reconjugated and mixed with newly synthesized bile acids to be re-secreted into bile. This process of intestinal deconjugation and hepatic reconjugation is a normal part of bile acid metabolism. This bacterial modification induces the epimerization of the C-7 hydroxy group of CDCA to form ursodeoxycholic acid (UDCA, 3α, 7β-dihydroxy-5β-cholanoate). UDCA is conjugated in the liver, circulates with the pool of primary bile acids, and normally constitutes less than 5% of the biliary bile acid pool [251–254].

Of note is that the secondary bile acids can be metabolized further by the liver or by the intestinal flora to form tertiary bile acids. This reaction, i.e., a process of re-epimerization of 3β-hydroxy bile acids in the liver, involves the reduction of 7-oxo-lithocholic acid to CDCA or its 7β-epimer to UDCA. Also, hepatic sulfation, hydroxylation, and glucuronidation of LCA form sulfolithocholic acid. Modification of LCA with sulfate or glucuronide can prevent its active uptake by the bile acid transporter ASBT and its passive absorption in the ileum. As a result, the conjugated LCA is rapidly lost from the circulating pool of bile acids. Because unmodified LCA is remarkably toxic to the liver, sulfation and hydroxylation of LCA have a protective effect on the liver function [178, 255–259]. Furthermore, the molecular mechanisms responsible for inducing these tertiary modifications have been elucidated. These processes involve the induction of cytosolic sulfotransferases and the cytochrome P4503A (CYP3A) by the orphan nuclear receptors pregnane X receptor (PXR) and constitutive androsterone receptor (CAR). LCA and other xenobiotic inducers can bind and directly activate PXR in the hepatocyte. The activated nuclear receptors then induce the expression of modification enzymes that confer resistance to LCA toxicity. Animal models pretreated with PXR ligands, such as the antipruritic agent rifampin, are protected against LCA-induced liver damage. In contrast, animal models lacking PXR and CAR are extremely sensitive to LCA-induced liver damage, which is characterized by necrotic foci in the liver and elevations of serum alanine aminotransferase levels. Because of the anticholestatic actions of these nuclear receptor pathways, they may be a promising target for treating patients with liver disease.

REGULATION OF BILE ACID BIOSYNTHESIS

Hepatic bile acid biosynthesis is downregulated primarily by a negative feedback system that is determined by a group of nuclear receptors [260–266]. The rate-limiting enzyme for regulating bile acid synthesis is cholesterol 7α-hydroxylase (CYP7A1) and the downregulation of this enzyme is mediated by several different pathways, all of which converge on CYP7A1 [173, 212, 213, 218, 223, 261–264, 267–271]. Many animal experiments and human studies have provided direct evidence showing bile acid feedback inhibition of CYP7A1 [272–283]. It has been found that bile acid synthesis is reduced by about 50% by feeding hydrophobic bile acids and increased 5- to 20-fold by interruption of the enterohepatic circulation following ileal resection or administration of bile acid sequestrants. In addition, the CYP7A1 mRNA transcripts in the 3′-untranslated region are unusually long and have a very short half-life of about 30 minutes. Bile acids reduce CYP7A1 mRNA stability via the bile acid response elements located in the 3′-untranslated region. Thus, many factors such as bile acids, steroid hormones, inflammatory cytokines, and insulin can inhibit *CYP7A1* transcription through the 5′-upstream region of the promoter [283–293]. Numerous studies have found that the farnesoid X receptor (FXR) plays a critical role in the regulation of bile acid metabolism because it can significantly inhibit *CYP7A1*, *CYP8B1*, and *CYP27A1* transcription by complicated negative feedback mechanisms.

It is now well recognized that bile acid biosynthesis is determined by two modes of regulation and both of these regulatory mechanisms are signaled by the intracellular concentration of bile acids, which act on the nuclear receptor FXR (Figure 5.5): (i) In the liver, bile acids stimulate FXR, activating an atypical nuclear receptor small heterodimer partner (SHP), which subsequently inhibits two major nuclear receptors, liver-related homolog-1 (LRH-1) and hepatocyte nuclear factor 4α (HNF4α), and induces an inhibitory effect on the transcription of CYP7A, the vital regulatory

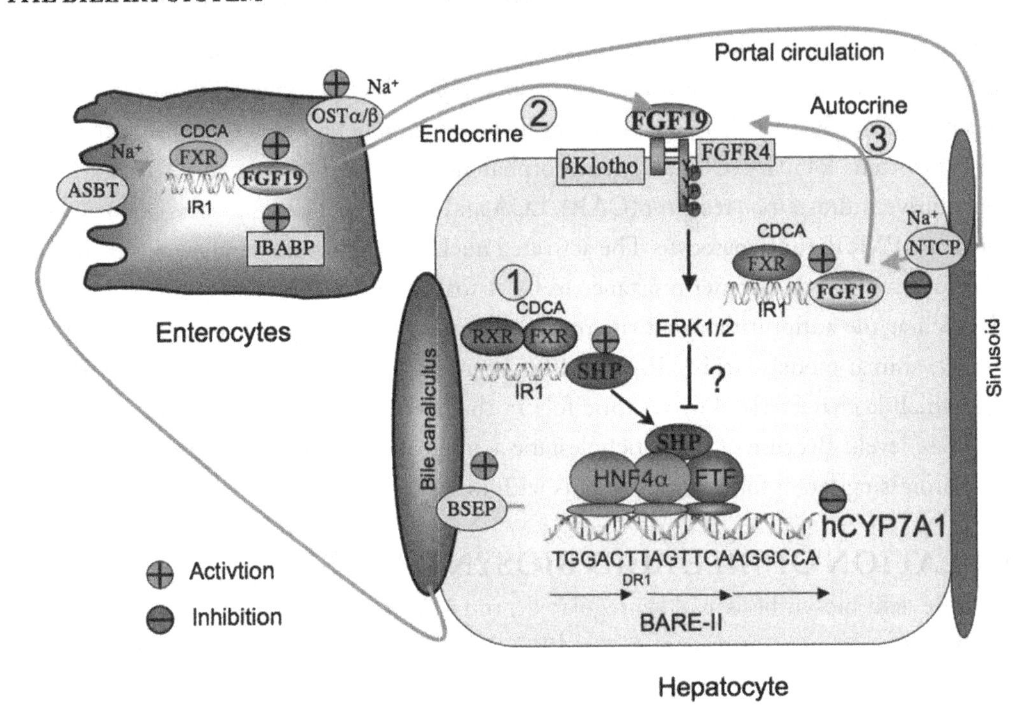

FIGURE 5.5: FXR-dependent mechanisms on the inhibition of cholesterol 7α-hydroxylase (CYP7A1) and the regulation of bile acid biosynthesis. **Pathway 1**: In the liver, bile acids activate the farnesoid X receptor (FXR), which up-regulates the expression of small heterodimer partner (SHP). SHP then inhibits the transactivation of *CYP7A1* through the liver-related homolog-1 (LRH-1) and hepatocyte nuclear factor 4α (HNF4α). **Pathway 2**: In the endocrine pathway, intestinal bile acids activate FXR, which stimulates the expression of fibroblast growth factor 19 (FGF19). Intestinal FGF19 is circulated to the liver to inhibit bile acid synthesis by activating hepatic FGF receptor 4 (FGFR4) signaling. **Pathway 3**: In the autocrine pathway, cholestatic bile acids may activate FXR and FGF19/FGFR4 signaling, which activates the MAPK/ERK1/2 pathway to inhibit *CYP7A1* transcription. However, this pathway may be an adaptive response to protect liver from cholestatic injury. Used with permission from Chiang JY. *Journal of Lipid Research*. 2009;50(10):1955–1966.

gene in bile acid synthesis. (ii) In the intestine, FXR activated by bile acids can promote a release of an intestinal peptide, fibroblast growth factor 19 (FGF19). As a result, intestinal FGF19 is circulated to the liver to inhibit bile acid synthesis by activating hepatic FGF receptor 4 (FGFR4) signaling. However, the inhibitory mechanism of the FXR/FGF19/FGFR4 signaling pathway on CYP7A1 transcription remains unknown. In addition, it has been found that bile acids may be able to stimulate a release of FGF19 from the hepatocyte, and the FGF19 autocrine pathway may exist

TABLE 5.4: The nuclear receptors and protein factors involved in the regulation of bile acid synthesis and enterohepatic cycling.

PROTEIN (GENE)	DESCRIPTION AND FUNCTION OF PROTEIN IN BILE ACID METABOLISM
FXR (*NR1H4*)	Bile acid activates nuclear receptor; regulation of bile acid synthesis, transport, and metabolism
HNF4α (*NR2A1*)	Nuclear receptor; positive regulator of cytochrome P-450 7A1 (CYP7A1) expression and hepatic bile acid synthesis
SHP (*NR0B2*)	Nuclear receptor; negative feedback regulation of hepatic bile acid synthesis by antagonizing HNF4α, LRH-1; regulation of bile acid transport and metabolism
LRH-1 (*NR5A2*)	Nuclear receptor; positive regulator of CYP7A1 expression and hepatic bile acid synthesis
PXR (*NR1I2*)	Bile acid and xenobiotic-activated nuclear receptor involved in detoxification of secondary bile acids
VDR (*NR1I1*)	Vitamin D and bile acid-activated nuclear receptor; involved in detoxification of LCA
FGFR4 (*FGFR4*)	Membrane receptor; negative feedback regulation of CYP7A1 and hepatic bile acid synthesis
β-klotho (*KLB*)	Membrane co-receptor associated with FGFR4; confers liver specificity to FGF4-FGF19 pathway; negative feedback regulation of CYP7A1 and hepatic bile acid synthesis
FGF19 (*FGF19*)	Protein growth factor; secreted by intestine in response to bile acids; regulates hepatic bile acids synthesis via FGFR4:β-klotho

Abbreviations: FXR, farnesoid X receptor; FGF19, fibroblast growth factor 19; FGFR4, fibroblast growth factor receptor 4; HNF4α, hepatocyte nuclear factor 4 alpha; LCA, lithocholic acid; LRH-1, liver receptor homolog 1; PXR, pregnane X receptor; SHP, small heterodimer partner; VDR, vitamin D receptor.

Dawson PA. Bile Secretion and the Enterohepatic Circulation in *Sleisenger and Fordtran's Gastrointestinal and Liver Disease*, the 9th Edition. Editors by Feldman M, Brandt L, and Friedman L. Elsevier Saunders. 2010; p. 1078.

in the human liver. Table 5.4 lists the nuclear receptors and protein factors that are involved in the regulation of bile acid synthesis and the enterohepatic circulation of bile acids in humans.

INHIBITORY MECHANISMS OF FXR SIGNALING ON CYP7A1

The regulation of bile acid biosynthesis by CYP7A1 at a molecular level has been studied extensively [262, 269, 284–286, 294–297]. It has been identified that the CYP7A1 is the FXR target gene [298–300]. The mechanism underlying FXR signaling in the inhibition of *CYP7A1* transcription and bile acid synthesis has been investigated [301–303]. These studies have revealed that there are two FXR-dependent mechanisms for bile acid inhibition of *CYP7A1* gene transcription. In the liver, FXR can inhibit CYP7A1 via the SHP pathway [223, 265, 274, 304–307]. In the intestine, FXR activated by bile acids could stimulate a release of FGF19 that inhibits CYP7A1 by triggering hepatic FGFR4 signaling [308–315].

The FXR/SHP Pathway

As illustrated in Figure 5.5, bile acids bind to FXR in the nucleus of hepatocyte. However, because the FXR binding site is not present in the *CYP7A1* promoter, *FXR* cannot directly inhibit *CYP7A1*. An FXR-RXR heterodimer acts to stimulate the synthesis of an inhibitory protein SHP, an atypical orphan nuclear receptor that has no DNA-binding domain and is a common transcriptional repressor of nuclear receptors. After SHP is activated by FXR, it inhibits the transactivating activity of LRH-1, thus inducing the inhibition of CYP7A1 transcription [298]. Also, SHP stimulated by FXR in turn can displace a promoter factor (HNF4α) from the promoter of the *CYP7A1* gene [316–319]. As a result, SHP interacts with HNF4α to prevent HNF4α interaction with PGC-1α, which can also inhibit *CYP7A1* and *CYP8B1* transcription. Of note is that the LRH-1 and HNF4α binding sites overlap in the *CYP7A1* and *CYP8B1* promoters. Because LRH-1 is a weak transcription factor and may compete with HNF4α for binding to the bile acid response element (BARE), it could directly inhibit *CYP7A1* and *CYP8B1* transcription as well [298, 316, 320].

The FXR/FGF19/FGFR4 Pathway

An additional essential negative regulatory factor is the fibroblast growth factor 19 (FGF19), a peptide secreted by the ileal enterocyte [308, 311, 313, 321]. FGF19 is also an FXR target gene as revealed by a microarray analysis of human primary hepatocytes treated with an FXR agonist GW4064. This peptide is able to trigger FGFR4 on the basolateral membrane of hepatocyte. After activating receptor tyrosine kinase FGFR4 signaling in the hepatocyte, FGF19 eventually induces a synergized suppression of *CYP7A1* transcription via a JNK-mediated pathway (Figure 5.5). Of note is that FGF19 activation of FGFR4 signaling requires β-Klotho, a membrane-bound glycosidase coexpressed with FGFR4 in the liver [322]. FGF19 has been detected in the plasma of humans [323–325]. Interestingly, FGF19 concentrations in the plasma show a diurnal variation that peaks 90–120 minutes after a postprandial rise in both bile acids and 7α-hydroxy-4-cholesten-3-one concentrations in the plasma [323]. Also, administration of CDCA increases FGF19 expression, whereas a bile acid-binding resin cholestyramine reduces FGF19 expression in the intestine. These observations support the concept that intestinal FGF19 could play an important role in inhibiting bile acid biosynthesis through the FXR/FGF19/FGFR4 pathway. Therefore, FGF19 could function as an enterohepatic signal to regulate bile acid synthesis in the liver [312, 326, 327].

Overall, the result of these complex regulatory circuits is that bile acid biosynthesis is autoregulated by bile acids themselves in a negative feedback manner.

FXR-INDEPENDENT BILE ACID INHIBITION OF CYP7A1

Many studies also suggested that bile acid inhibition of CYP7A1 may be determined by several FXR-independent mechanisms (Figure 5.6). The secondary bile acid LCA is a ligand of pregnane X receptor (PXR) and vitamin D receptor (VDR). These two receptors bind to the bile acid response element (BARE)-I sequence in the *CYP7A1* promoter and inhibit *CYP7A1* promoter activity [328–334]. PXR and VDR inhibit *CYP7A1* transcription by blocking HNF4α recruitment of PGC-1α to CYP7A1 chromatin. Because PXR induces *CYP27A1* expression in the enterocyte but not in the hepatocyte, it suggests that PXR may play a role in the regulation of *CYP7A1* gene transcription by an indirect mechanism. There are several mechanisms for the LCA-activated VDR inhibition of *CYP7A1* gene transcription [335–337]: (i) VDR may compete with HNF4α for coactivators; (ii) VDR may recruit corepressors to the *CYP7A1* promoter; (iii) VDR and HNF4α may compete for binding to the DR1 motif; and (iv) VDR may blockade HNF4α interaction with PGC-1α. In addition, the constitutive androstane receptor (CAR) inhibits *CYP7A1* transcription by binding to the DR1 motif and competing with HNF4α for coactivators PGC-1α.

In the hepatocyte, insulin signaling phosphorylates and activates the insulin receptor leading to the activation of insulin receptor substrates, PI3K and AKT, which phosphorylate FoxO1 and inhibit *CYP7A1* transcription. Bile acids also activate epidermal growth factor receptor and the Raf-1/MEK/ERK signaling pathway to inhibit *CYP7A1* transcription. Bile acid-activated protein kinase C phosphorylates cJun to inhibit *CYP7A1* transcription.

During liver injury and regeneration, hepatocyte growth factor (HGF) released from hepatic stellate cells could inhibit *CYP7A1* transcription and bile acid synthesis by stimulating the HGF receptor cMet and MAPK pathways [338–341]. During the endotoxin-induced cholestasis, lipopolysaccharides stimulate the release of TNFα and IL-1β from the Kupffer cells by activating the Toll-like receptor 4. TNFα and IL-1β may inhibit *CYP7A1* transcription by activating the TNF receptor and the MAPK/JNK pathway in the hepatocyte [342–344]. JNK may inhibit *CYP7A1*

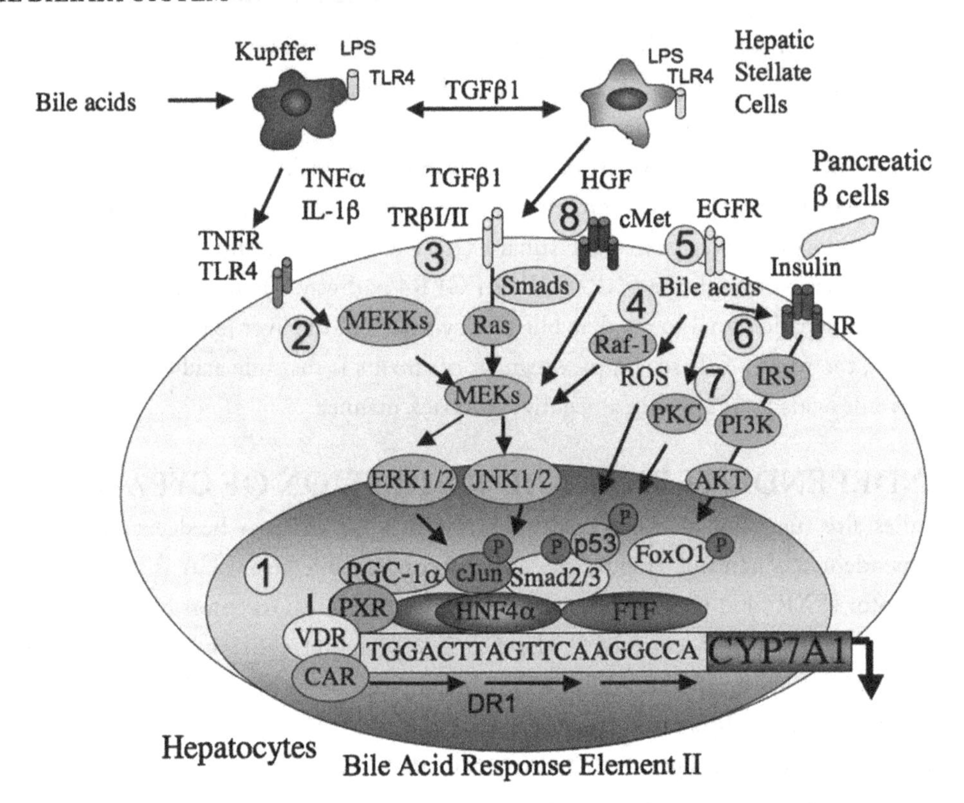

FIGURE 5.6: FXR-independent and bile acid-activated cell signaling pathways in regulation of *CYP7A1* transcription. All these signaling pathways as indicated in number may converge to regulate CYP7A1 chromatin structure by the epigenetic mechanism. Used with permission from Chiang JY. *Journal of Lipid Research*. 2009;50(10):1955–1966.

and *CYP8B1* transcription and bile acid synthesis by phosphorylating cJun and HNF4α [326, 345–350]. TGFβ-1 secreted from Kupffer cells activates its receptor TRβII and the SMAD signaling pathway in hepatocytes. SMAD3 enters the nucleus of hepatocyte and works with HDACs and mSin3A to inhibit HNF4α activation of *CYP7A1* transcription. A tumor suppressor p53 interacts with HNF4α and inhibits HNF4α activity. These alterations may inhibit *CYP7A1* transcription. Taken together, all these cell signaling pathways could induce a FXR-independent bile acid inhibition of *CYP7A1* transcription, which may protect against bile acid toxicity during cholestasis and liver injury.

• • • •

CHAPTER 6

The Enterohepatic Circulation of Bile Acids

PHYSIOLOGY OF THE ENTEROHEPATIC CIRCULATION

The enterohepatic circulation of bile acids is defined as the recycling of biliary bile acids from the liver, where they are synthesized and secreted into bile, to the small intestine, where it facilitates the absorption of dietary fat and other lipid-soluble substances, and back to the liver [229, 351–353]. The major anatomical components of the enterohepatic circulation include the liver, biliary tract, gallbladder, small intestine, and portal venous circulation. Also, the colon, systemic circulation, and kidney play minor roles in the enterohepatic circulation of bile acids. The sequential processes of hepatic bile acid secretion into the biliary tract, concentrating and storage by the gallbladder, active and efficient absorption by the small intestine, and recapture by the liver are the essence of the enterhepatic circulation. Because more than 95% of the secreted bile acids are actively reabsorbed by an apical sodium-dependent bile acid transporter (ASBT) in the distal small intestine [354], efficient conservation of these compounds results in the accumulation of a large mass of bile acids that circulates between the liver and small intestine. As a result, less than 5% of the cycling bile acids escape intestinal reabsorption and are eliminated in the feces. Therefore, in the normal physiological state, the quantity of bile acids synthesized in the liver is equal to the amounts of bile acids lost in the feces every day, which maintains a constant bile acid pool. In a healthy adult, small amounts (0.2 to 0.6 g/day) of bile acids are lost in the feces. Thus, less than 5% of the bile acids present in hepatic bile are newly synthesized. This effective recycling and conservation mechanism largely restricts bile acids to the hepatobiliary and intestinal compartments. Furthermore, bile acids returning to the hepatocyte can regulate their own synthesis from cholesterol by a negative feedback mechanism, and thus, may indirectly regulate *de novo* cholesterol synthesis [262]. Figure 6.1 illustrates the enterohepatic circulation of bile acids in healthy humans.

During the fasting period, bile acids secreted by the hepatocytes enter the biliary tract and move down to the gallbladder, where they are concentrated. The concentration of bile acids in gallbladder bile is approximately 10-fold higher compared with that of hepatic bile. Because most of the bile acids are sequestered in the gallbladder after an overnight fast, their concentrations are quite

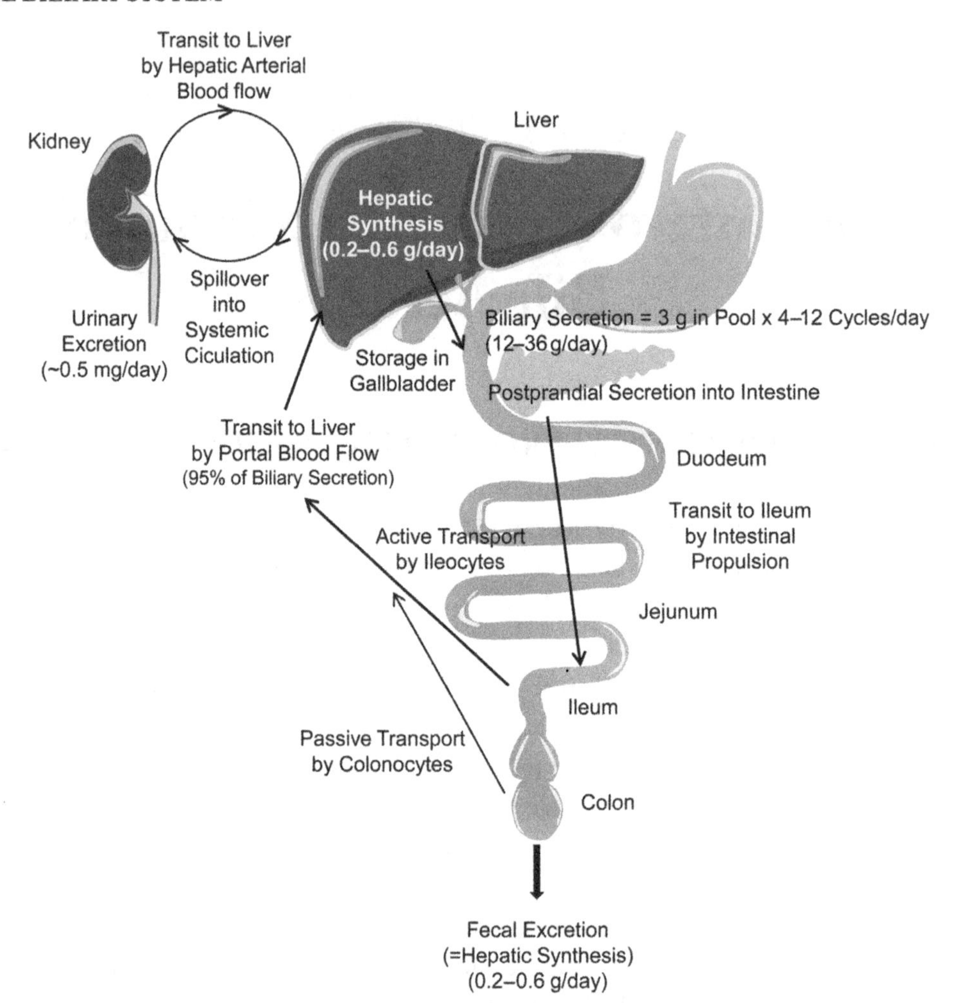

FIGURE 6.1: Schematic depiction of the enterohepatic circulation of bile acids in humans. The distinct kinetic compartments involved in the enterohepatic circulation include the liver, biliary tract, gallbladder, small intestine, large intestine, kidney, portal circulation, and systematic circulation. Because intestinal conservation of bile acids is highly efficient, approximately 95% of circulating bile acids are absorbed on each pass through the intestine and are returned to the liver. In a healthy adult, small amounts (0.2 to 0.6 g/day) of bile acids are lost in the feces, which are replaced by the newly synthesized bile acids in the liver. So, a constant bile acid pool is maintained.

low in the small intestine, portal vein, plasma, and liver. When a meal is ingested, dietary fat and protein trigger a release of cholecystokinin (CCK) from the proximal small intestine [65]. Because CCK stimulates gallbladder contraction and acts on the biliary tree to relax the sphincter of Oddi, bile acids that are stored in the gallbladder are slowly discharged into the duodenum. During the digestion of a large meal, the gallbladder remains contracted, and bile acids secreted by the liver bypass the gallbladder and directly enter the proximal small intestine [355]. During this consumption period, the bile acid concentration in the small intestine is approximately 5 to 10 mmol/L. When the intraluminal concentration of bile acids is greater than 1.5 mmol/L, they can form a large amount of simple micelles. Obviously, other biliary lipid components such as cholesterol and phospholipids are also emptied from the gallbladder into the small intestine. Bile acids, together with these biliary lipids, can form a large number of mixed micelles [356–358]. In the intestinal lumen, these micelles greatly facilitate the digestion and absorption of dietary fat by the small intestine through solubilizing the hydrolytic compounds, triggering the action of pancreatic lipase on the hydrolysis of triglyceride, and delivering lipids to the apical membrane of enterocytes [82]. Because the enterohepatic circulation of bile acids leads to the accumulation of a large amount of bile acids in the small intestinal lumen, these detergent molecules can be repeatedly used during the digestion and absorption of fat for each meal throughout the day.

BILE ACID POOL AND KINETICS

Because the cycling of bile acids between the liver and the intestine is referred to as the enterohepatic circulation, the total amount of bile acids in the enterohepatic circulation is defined as a circulating bile acid pool [229, 352, 359]. Table 6.1 shows pool size and kinetics of individual bile acids in healthy subjects. Because of the complexity of the methodologies involved, the bile acid pool is measured by an isotope dilution technique. The kinetics for bile acid turnover in humans can be derived from measurements of bile acid pool size and hepatic secretion rates by intestinal perfusion methods, together with an adequate stimulus for gallbladder contraction [360–362]. Bile acid synthesis in the liver can be determined by balance techniques or by modified Lindstedt methods [363]. From these measurements, the recycling frequency of the pool and intestinal absorption efficiency of bile acids can be calculated. Furthermore, division of hepatic bile acid secretion by the pool size gives the recycling frequency of the bile acid pool. In a healthy adult, the enterohepatic circulation maintains a bile acid pool size of approximately 2 to 4 g/kg body weight. The bile acid pool cycles two to three times per meal, resulting in 6 to 10 cycles per day, and the intestine may reabsorb between 10 and 30 g of bile acid per day. As shown in Figure 6.1, approximately 0.2 to 0.6 g of bile acids escapes intestinal reabsorption and is eliminated in the feces each day. Thus, the size of the bile acid pool is the dependent variable, and can be regulated by total bile acid synthesis

TABLE 6.1: Pool size and kinetics of individual bile acids in healthy subjects.

BILE ACID	POOL SIZE (mg)	FRACTIONAL TURNOVER RATE (day^{-1})	DAILY SYNTHESIS (mg/day)	DAILY INPUT FROM PRIMARY BILE ACIDS (mg/day)
Cholate	500–1,500	0.2–0.5	120–400	–
Deoxycholate	200–800	0.1–0.4	–	40–200
Chenodeoxycholate	500–1,200	0.2–0.4	100–250	–
Lithocholate	50–150	0.8–1.0	–	50–100
Total	1,300–3,650	–	220–650	90–300

Used with permission from Dawson PA. Bile Secretion and the Enterohepatic Circulation. In: *Sleisenger and Fordtran's Gastrointestinal and Liver Disease.* Editors by Feldman M, Brandt L, and Friedman L. the 9th Edition. Elsevier Saunders. 2010; p. 1077.

in the liver, total hepatic secretion rate of biliary bile acids, intestinal absorption efficiency of bile acids, and recycling frequency of the bile acid pool.

Figure 6.2 shows a method for measuring enterohepatic circulatory dynamics in animal models such as mice, rats, rabbits, hamsters, and monkeys with bile fistula in which hepatic bile samples are continuously collected to determine hepatic bile acid output and the bile acid pool size [359, 363–367]. This "washout" technique provides a precise method for the direct measurement of bile acid pool size, as well as basal and compensated bile acid synthesis rates.

DRIVING FORCES OF THE ENTEROHEPATIC CIRCULATION

Active secretion of bile acids across the canalicular membrane of hepatocyte into bile by the bile acid export pump ABCB11 is the primary driving force of the enterohepatic circulation, which is the rate-limiting step in the overall transport of bile acids from the liver to bile [368–372]. Also, the presence of an ileal active transport system significantly improves the intestinal absorption efficiency of bile acids [354]. Because hepatic secretion of bile acids induces bile flow, maintaining the enterohepatic circulation also results in continuous bile formation. The dissociation of bile acid synthesis from intestinal delivery is also promoted by the presence of the gallbladder, because the availability of a concentrative storage reservoir induces bile acids to be delivered in a high concentration and determines how much amounts of bile acids are excreted to the duodenum. The ileal bile acid transporter such as ASBT and the gallbladder are complementary rather than redundant,

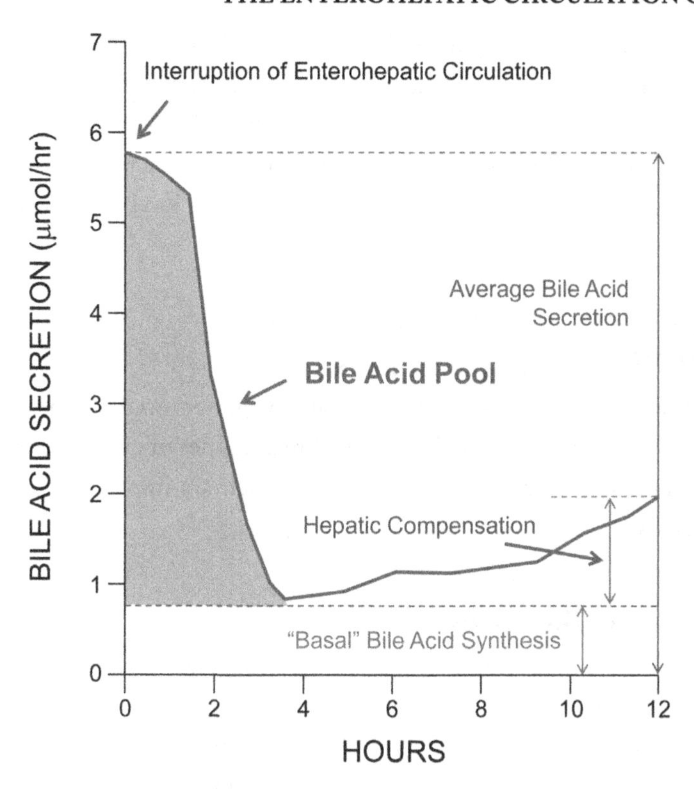

FIGURE 6.2: Measurement of bile acid pool by a "washout" technique. This diagram shows washout curves of hepatic bile acid output in animals after cannulation of the common bile duct (100% interruption of the enterohepatic circulation). Bile acid secretion rate falls over the subsequent 2 hours and reaches a low point at 3 hours. The shaded area shows the direct measurement of bile acid pool sizes. The low point (dashed line) at 3 hours represents "basal" bile acid synthesis in the liver. When the bile acid secretion increases again, this indicates that newly synthesized bile acids are appearing. Subsequently, bile acid secretion rate reaches a new steady state due to hepatic compensation.

both of them working together to conserve bile acids in the body [355]. In the presence of a gallbladder but the absence of an active ileal bile acid transporter, the secreted bile acids would not be reabsorbed efficiently [352]. Emptying of the gallbladder bile would necessarily be followed by a refractory period during which the bile acid supply would not be sufficient to help the digestion and absorption of dietary fat. The refractory period would last until hepatic synthesis could restore the bile acid pool. The existence of an ileal bile acid transporter and an enterohepatic circulation results in the repeated use of the bile acid pool during the consumption of a single meal. Moreover, a very small amount of bile acids is absorbed by the proximal small intestine. In contrast, most of the bile acid molecules are absorbed by the lower third of the small intestine where a high-affinity transporter ASBT on the apical membrane of the ileocytes (ileal enterocytes) actively absorbs bile acids. After bile acids are transferred into the portal blood, they are carried to the liver via the portal vein. In the liver, bile acids are taken up again by a high-affinity binding transporter on the sinusoidal membrane of hepatocyte. Because of the efficiency of the two active transport systems on the ileocyte and the hepatocyte, there is a very low plasma bile acid level (less than 3-5 μM) and few spillage of bile acids into urine.

KINETICS AND REGULATION OF THE ENTEROHEPATIC CIRCULATION

Table 6.2 lists the major hepatic, biliary, and intestinal transporters on the sinusoidal and canalicular membranes of hepatocytes, the apical and basolateral membranes of cholangiocytes, and the apical and basolateral membranes of ileal enterocytes, all of which are important for the enterohepatic circulation of bile acids.

The Hepatic Uptake of Bile Acids

Bile acids return to the liver from the intestine via the portal vein. In portal blood, trihydroxy bile acids are bound to albumin with moderate affinity (50–80%), and dihydroxy bile acids and lithocholic acid (LCA) with greater affinity (more than 95%). The majority of bile acids returning to the liver are in conjugated form, while a small portion is in unconjugated form. During late pregnancy, biosynthesized conjugated bile acids are transferred from the fetus to the mother across the placenta. In contrast, unconjugated dihydroxy bile acids are membrane-permeable and may transfer passively from mother to fetus or vice versa.

Hepatic uptake of bile acids is highly efficient (50–90%) and independent of load. The $T_{1/2}$ of plasma disappearance for all bile acids is less than five minutes in humans. The efficiency of uptake depends on bile acid structure and the state of conjugation. Furthermore, the small amount of bile acids that enters the glomerular filtrate is reabsorbed by ASBT in the proximal tubule of kidneys. As a result, loss of bile acids in the urine is negligible.

Hepatic Sinusoidal Sodium-Dependent Bile Acid Uptake

Because there is a 5- to 10-fold concentration gradient between the portal blood and hepatocyte cytosol, hepatic uptake of bile acids must overcome an unfavorable electrochemical ion gradient [369, 373]. So, the uptake of conjugated bile acids at the sinusoidal membrane of hepatocytes is mediated mainly by an active absorption mechanism through the sodium-taurocholate cotransporting polypeptide (NTCP) (Figure 6.3) [374–378]. This hepatic sinusoidal sodium-dependent

TABLE 6.2: Function of transport proteins involved in bile formation and the enterohepatic circulation of bile acids.

TRANSPORTER (GENE)	LOCATION	FUNCTION
Hepatocyte	***Bile acid-dependent bile flow***	
NCTP (*SLC10A1*)	Basolateral membrane	Na^+-dependent bile acid and xenobiotic uptake
OATP1B1 (*SLCO1B1*)	Basolateral membrane	Na^+-independent bile acid and xenobiotic uptake
OATP1B3 (*SLCO1B3*)	Basolateral membrane	Na^+-independent bile acid and xenobiotic uptake
Na^+, K^+-ATPase	Basolateral membrane	Secretion of 2 Na^+ in exchange for 3 K^+
BSEP (*ABCB11*)	Canalicular membrane	ATP-dependent bile acid export
MDR3 (*ABCB4*)	Canalicular membrane	ATP-dependent phosphatidylcholine export
ABCG5	Canalicular membrane	ATP-dependent sterol export
ABCG8	Canalicular membrane	ATP-dependent sterol export
NPC1L1	Canalicular membrane	Sterol import
FIC1 (*ATP8B1*)	Canalicular membrane	ATP-dependent aminophospholipid flipping
	Bile acid-independent bile flow	
MRP2 (ABCB2)	Canalicular membrane	ATP-dependent transport of glucuronide, glutathione, and sulfate conjugates
OATP (1B1, 1B3, 2B1)	Basolateral membrane	Na^+-independent transport of organic anions, cations, and neutral steroids

TABLE 6.2: *(continued)*

TRANSPORTER (GENE)	LOCATION	FUNCTION
	Sinusoidal bile acid export	
MRP3 (*ABCC3*)	Basolateral membrane	ATP-dependent export of bile acids and glucuronide conjugates
MRP4 (*ABCC4*)	Basolateral membrane	ATP-dependent export of glutathione and bile acids
OSTα /β	Basolateral membrane	Bile acid export
Cholangiocyte	***Ductular secretion***	
Aquaporin 1 (*AQP1*)	Apical membrane	Water transport
Aquaporin 4 (*AQP4*)	Basolateral membrane	Water transport
AE2 (*SLC4A2*)	Apical membrane	HCO_3^- secretion in exchange for Cl^-
CFTR (*ABCC7*)	Apical membrane	Cl^- secretion
ASBT (*SLC10A2*)	Apical membrane	Bile acid uptake (cholehepatic shunt)
Ileal enterocyte		
ASBT *(SLC10A2)*	Apical membrane	Na^+-dependent bile acid uptake
NPCL1L1	Apical membrane	Sterol import
OSTα /β	Basolateral membrane	Bile acid export
MRP3 *(ABCC3)*	Basolateral membrane	Bile acid export

ABC, ATP-binding cassette; AE2, chloride-bicarbonate anion exchanger isoform 2; ASBT, apical sodium-dependent bile acid transporter; BSEP, bile salt export pump; CFTR, cystic fibrosis transmembrane regulator; FIC1, P-type ATPase mutated in progressive familial intrahepatic cholestatis type 1; MDR, multidrug resistance protein; MRP, multidrug resistance-associated protein; NPC1L1, Niemann–Pick C1 Like 1; NTCP, sodium taurocholate cotransporting polypeptide; OATP, organic anion transporting polypeptide; OST, organic solute transporter; SLC, solute carrier.

Used with permission from Dawson PA. Bile Secretion and the Enterohepatic Circulation. In: *Sleisenger and Fordtran's Gastrointestinal and Liver Disease.* Editors by Feldman M, Brandt L, and Friedman L. the 9th Edition. Elsevier Saunders. 2010; p. 1083.

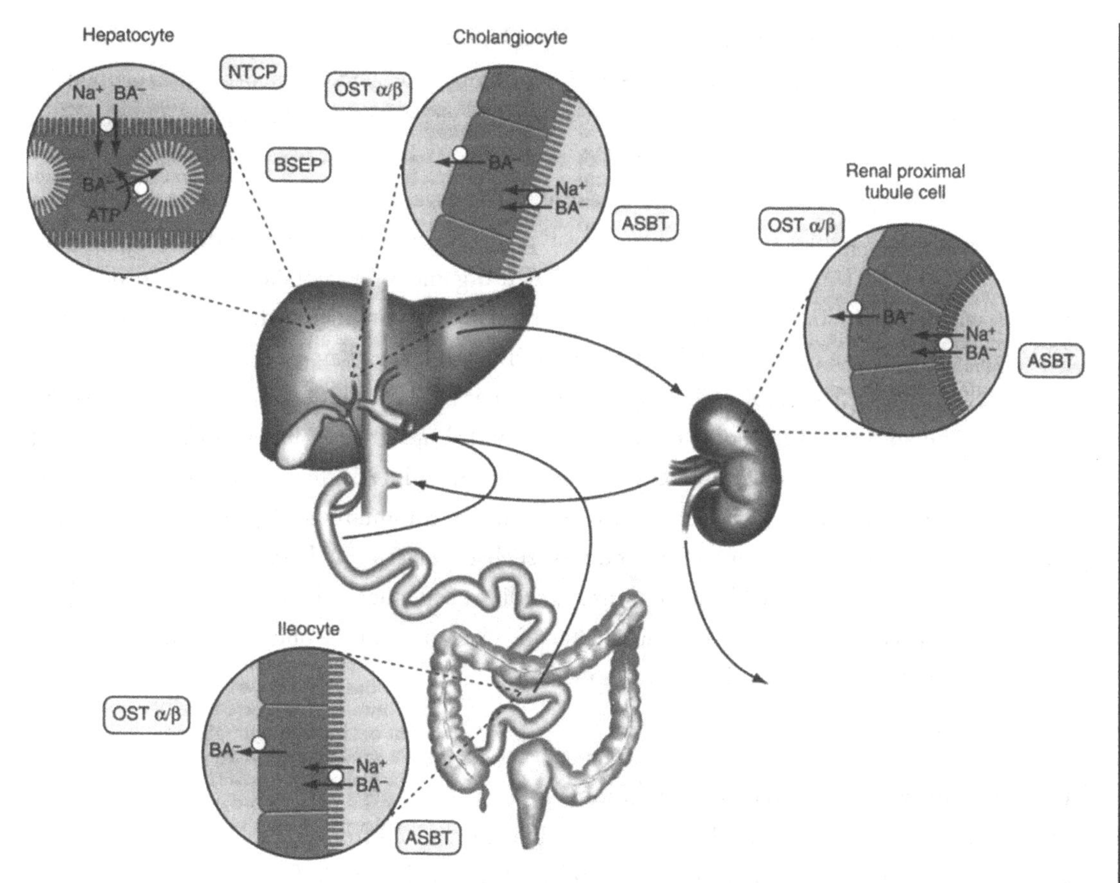

FIGURE 6.3: This diagram shows the individual transport proteins responsible for bile acid transport across the apical or the basolateral membrane of hepatocytes, cholangiocytes, ileocytes (ileal enterocytes), and renal proximal tubule during the enterohepatic circulation of bile acids. Abbreviations: ASBT, apical sodium-dependent bile acid transporter; BA, bile acid; BSEP (ABCB11), bile salt export pump; NTCP, sodium-taurocholate cotransporting polypeptide; OST, organic solute transporter. Used with permission from Mosely RH. Bile Secretion and Cholestasis. In: *Liver and Biliary Disease.* Editor: Kaplowitz N., 2nd Edition. Williams & Wilkins, Philadelphia. 1996; p. 194.

bile acid transporter plays a crucial role in active uptake of conjugated bile acids. In addition, the driving force for the sodium-dependent uptake is generated by the basolateral Na^+/K^+-ATPase that maintains the prevailing out-to-in sodium gradient. In contrast to conjugated bile acids, sodium-dependent uptake accounts for less than one half of the uptake of unconjugated bile acids such as cholic acid (CA) and ursodeoxycholic acid (UDCA).

Hepatic Sinusoidal Sodium-Independent Bile Acid Uptake

Although all the properties of hepatic sodium-dependent bile acid transport are explained by NTCP, sodium-independent bile acid transport plays a critical role in the uptake of bile acids. Especially, unconjugated bile acids such as CA and UDCA are taken up predominantly by hepatic sodium-independent transport systems through the organic anion transporting polypeptide (OATP) transporter family [379–381]. The OATP genes encode 12 potential transmembrane domain proteins. The driving force for OATP-mediated organic anion uptake appears to be anion exchange, with coupling of bile acid uptake to bicarbonate or, more likely, glutathione efflux. The OATP-type transporters constitute a large gene family with more than 36 members identified in human, mouse, and rat tissues [382]. It has been found that OATB1B1, OATB1B3, and OATB2B1 function in human liver to transport organic anions and drugs. These human OATPs transport bilirubin glucuronides, steroid metabolites (such as estradiol-17β-glucuronide and estrone-3-sulfate), arachidonic acid products (such as prostaglandin E_2, thromboxane B_2, and leukotriene C_4), bromosulphophthalein, and a wide variety of drugs (such as pravastatin, digoxin, and fexofenadine). Because the majority of hepatic bile acid uptake is sodium dependent, the major physiological role of these broad-specificity solute carriers may be hepatic clearance of non-bile acid substrates such as endogenous amphipathic metabolites and xenobiotics.

The Canalicular Transport of Bile Acids

Approximately 95% of the bile acids secreted into bile are derived from the recirculating pool and the newly-synthesized bile acids in the liver provide the remaining 5% for biliary secretion. Bile acids across the canalicular membrane of hepatocytes are the rate-limiting step for their transport from blood into bile, which is a transporter-mediated process important for bile formation. Because the concentrations of bile acids are more than 1,000-fold higher in the bile canaliculus than within the hepatocytes, an ATP-dependent bile acid transport system is responsible for an active transport of bile acids across the canalicular membrane. It has been found that a member of the ATP-binding cassette (ABC) transporter superfamily is the canalicular bile acid transporter named the bile acid export pump (BSEP, gene symbol *ABCB11*). Functional expression and characterization studies have revealed that ABCB11 can efficiently transport conjugated bile acids for biliary secretion. ABCB11 is a 160-kDa protein and is located on chromosome 2q24 in humans. The human *ABCB11* gene locus is linked to progressive familial intrahepatic homeostasis type 2 (PFIC2), a hepatic disorder characterized by an absence of ABCB11 on the canalicular membrane of hepatocytes and a striking lack of bile acids in bile [383–385]. Clinical studies have found that mutations in the *ABCB11* gene are associated with a very low level of bile acid secretion and severe cholestasis in PFIC2 patients. These studies further confirmed the role of ABCB11 as the major canalicular bile acid transporter in biliary secretion [386]. In addition, deletion of the *Abcb11* gene results in a significant deficiency in hepatic secretion of biliary bile acids in mice [101, 371, 372, 387].

The Role of Gallbladder in the Enterohepatic Circulation

The gallbladder is the final recipient of bile produced by the liver. Its major physiological functions include the storage of bile during the interdigestive period, the conversion of dilute hepatic bile into concentrated gallbladder bile, the delivery of bile into the intestine, and the acidification of bile. During the interdigestive period, the gallbladder concentrates bile about 5- to 10-fold by the coupled active absorption of cations (mainly sodium) and chloride/bicarbonate ions with passive movement of water. So, hepatic bile is concentrated to approximately 10% to 20% of its original volume in the gallbladder. Because of a large amount of absorption of water and electrolytes by the gallbladder epithelial cells, the process leads to a substantial increase in the concentration of bile acids. The progressive concentration of bile is associated with mixed micelle formation. The concentrated gallbladder bile with micelles greatly aids not only the absorption of fat, cholesterol, fat-soluble vitamins, and other nutrients, but also the digestion of fat. During food consumption, the small intestine releases CCK, an active polypeptide with the principal function of stimulating gallbladder contraction [355]. A good correlation between plasma concentration of CCK and gallbladder contraction has been observed. Contraction of the gallbladder, coupled with the synchronous relaxation of the sphincter of Oddi, results in the emptying of gallbladder bile into the duodenum, with as much as 80% of its contents being discharged into the duodenum. Figure 6.4 shows that gallbladder emptying plays a crucial role in determining the kinetics of enterohepatic circulation of bile acids [388–390].

Interdigestively, the gallbladder remains relaxed, which is mediated mainly by fibroblast growth factor 19 (FGF19) that is secreted by the ileocytes, i.e., the ileal enterocytes [391]. Also, the gallbladder relaxation is partially due to diminished circulating levels of CCK and the absence of neurohumoral stimulation [392]. The intestinal FGF19 is an FXR target gene, and its concentrations in the plasma show a diurnal variation that peaks 90–120 minutes after a postprandial rise. The deletion of the *Fgf15* gene (the mouse homolog of *FGF19*) induces impaired gallbladder refilling in mice [391]. This "active" gallbladder relaxation draws hepatic bile into its lumen, which results in gallbladder refilling [393–395]. The secretory pressure of the liver and the sphincter resistance is coordinated to account for gallbladder filling and emptying [112].

Using a dual marker method for simultaneous measurement of hepatic bile secretion and gallbladder bile emptying, it has been observed that most hepatic bile is diverted into the gallbladder postprandially and during fasting [396]. Studies in baboons and humans also found that about one-half the hepatic bile enters the gallbladder for concentration and storage, while the other half

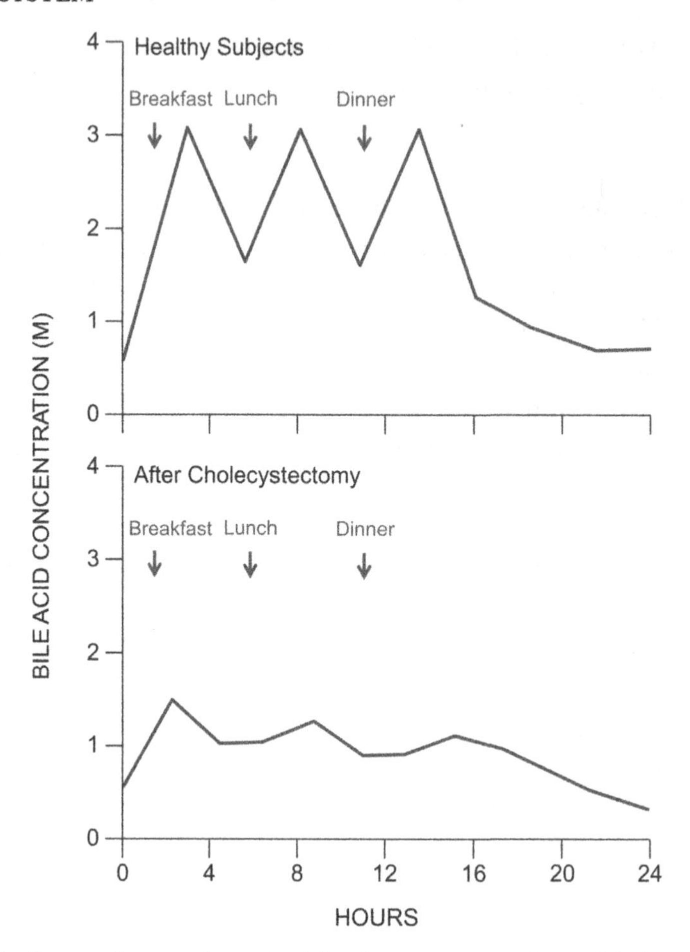

FIGURE 6.4: This diagram shows a change in daily plasma concentrations of cholate conjugates detected by radioimmunoassay. Top panel displays a fluctuating enterohepatic circulation in healthy subjects. Bottom panel shows a continuous enterohepatic circulation after cholecystectomy in patients.

bypasses the gallbladder to enter the duodenum and undergoes continuous enterohepatic cycling [397]. Further studies found that in the interdigestive phase, the gallbladder undergoes intermittent partial emptying during phase II of the migrating myoelectric complex. Moreover, considering that only a fraction of bile enters the gallbladder, even in the presence of a normally functioning sphincter of Oddi, it would appear that enterohepatic circulation of bile acids never completely ceases [269]. It seems, therefore, that gallbladder contraction is the major factor influencing bile entering the duodenum during fasting. The gallbladder participates as a holding reservoir and mechanical pump in the enterohepatic circulation.

Of note, many mammals and birds have no gallbladder [398–400], and cholecystectomy does not impair the digestion and absorption of dietary fat in patients [39, 401–404]. These observations suggest that the storage and concentration of bile in the gallbladder between meals are not necessary for the digestion and absorption of dietary lipids, and the intestinal concentration of bile acids is adequate for the absorption of dietary lipids when there is no gallbladder. In addition, although deficient gallbladder contraction probably plays a minor role in fat absorption, it may be of importance in the pathogenesis of cholesterol gallstone formation.

The Role of the Intestine in the Maintenance of the Enterohepatic Circulation

It has been found that the efficient intestinal conservation of bile acids is mediated mainly by two mechanisms: the active transporter-mediated bile acid uptake and the passive diffusion absorption. As conjugated bile acids are membrane-impermeable, the dominant mechanism for the absorption of conjugated bile acids is active absorption by a transporter-mediated mechanism in humans and in most animal models such as mice and rats. Bile acids are actively transported across the apical membrane of ileal enterocytes by the apical sodium-dependent bile acid transporter (ASBT), whose synthesis is mediated by the gene *SLC10A2* [250, 354, 405–408]. Absence of a normal *SLC10A2* gene in humans or knockout of this gene in mice can cause profound intestinal bile acid malabsorption. At the basolateral membrane of ileocytes (ileal enterocytes), bile acid transport is mediated by a heterodimer of two proteins—the organic solute transporters α and β (OSTα/β) [379, 409–412].

Ileal Sodium-Dependent Bile Acid Uptake

After each meal, gallbladder contraction empties bile acids into the duodenum. When passing through the intestinal tract, a very small amount of the bile acids are reabsorbed in the proximal small intestine by passive diffusion, but approximately 95% of the bile acids are reabsorbed by the active transport system through ASBT in the distal small intestine (predominantly in the ileum). Both the sodium gradient maintained by the basolateral Na^+/K^+-ATPase and the negative intracellular potential provide the driving force for ASBT-mediated bile acid uptake in the intestine. The ASBT transports all major species of bile acids but favors trihydroxy (i.e., CA) over dihydroxy bile acids and conjugated over unconjugated species. Moreover, because glycine-conjugated dihydroxy bile acids are membrane-permeable, it is highly likely that passive absorption across the apical membrane of enterocytes occurs when duodenal content is acidic.

Bile acids largely exist in the conjugated form in bile; however, unconjugated bile acids can be formed by bacterial enzymes. Such deconjugation begins mainly in the ileum in humans. Moreover, deconjugation is rapidly completed when bile acids enter the large intestine. The resultant unconjugated bile acids have varying degrees of passive membrane permeability. A monohydroxy bile acid

LCA and probably most dihydroxy bile acids are highly membrane-permeable. It is likely that the magnitude of unconjugated bile acid absorption in the distal small intestine exceeds that of unconjugated bile acid absorption from the colon, based on modeling of the enterohepatic circulation of bile acids in humans. Although trihydroxy bile acids have much more limited membrane permeability, they are absorbed actively by ASBT in the distal small intestine. In the colon, deoxycholic acid (DCA) is reabsorbed by passive transport and recycled with CA and chenodeoxycholic acid (CDCA) to the liver.

Little is known about the intracellular transport of bile acids in the ileal enterocyte. It has been assumed that after entering the ileocytes, bile acids are transdiffused across the enterocyte to the basolateral membrane. The heterometric organic solute transporters OSTα/β are mainly responsible for bile acid export across the basolateral membrane of ileocytes and have an ability to export all the major bile acid species into portal blood circulation back to the liver. Also, this process may be mediated by an anion exchange process or an ATP-dependent transporter multidrug resistance-associated protein 3 (MRP3, gene symbol *ABCC3*) [270, 413–416].

The Role of Kidneys in the Enterohepatic Circulation

Renal Sodium-Dependent Bile Acid Uptake

It has been found that the sodium-bile acid cotransporters are all products of the same gene *SLC10A2* in the cholangiocyte, ileal enterocyte, and renal proximal tubule cell, and they share considerable amino acid identity and structural similarity [369, 406, 407, 417–421]. In the kidneys, ASBT also mediates bile acid uptake across the apical membrane of the renal proximal tubule cell. The bile acid transporters OSTα/β are involved in bile acid transport across the basolateral membrane of the renal proximal tubule cell to the renal peritubular capillaries, and then, these bile acids circulate back to the liver [379]. Basolateral bile acid export in the renal proximal tubule cells may be mediated by one of the ATP-dependent transporter C family, i.e., multidrug resistance-associated protein 3 (MRP3) or an anion exchange process.

• • • •

CHAPTER 7

Hepatic Secretion of Biliary Lipids and Bile Formation

STRUCTURE OF THE BILE SECRETORY APPARATUS

The hepatic secretion of bile begins in the bile canaliculus, the smallest branch of the biliary tree. The canaliculi are formed by a specialized membrane of adjacent apical poles of two hepatocytes, which form a meshwork of polygonal channels between hepatocytes. Bile then enters the small terminal channels (the canals of Hering), which have a basement membrane and are lined partly by hepatocytes and partly by cholangiocytes. The canals of Hering provide a conduit through which bile enters the larger perilobular or intralobular ducts. Moreover, the smallest biliary radicles are a small space surrounded by cuboidal epithelial cells of bile ducts. At the most proximal level, one or more ductular cells may share a canalicular lumen with a hepatocyte; gradually, the ductules become lined by two to four cuboidal epithelial cells as they approach the portal canal. Bile flows from the central lobular cells toward portal triads (from zone 3 to zone 1 of the liver acinus). The interlobular bile ducts are lined by a layer of cuboidal or columnar epithelium that displays a microvillar architecture on its luminal surface. Finally, these bile ducts carry bile away from the liver lobules.

SOURCE OF LIPIDS SECRETED IN BILE

Bile formation is an osmotic process; solutes are actively transported into the canaliculus by primary active transporters such as ABCG5/G8 for biliary cholesterol secretion, ABCB4 for biliary phospholipid secretion, and ABCB11 for biliary bile acid secretion [26, 27]. The most important solutes driving bile formation are bile acids. Bile formation serves three important functions [52]: First, it is a major route for the elimination of cholesterol from the body, either as unesterified cholesterol or as bile acids, the end productions of cholesterol degradation. Second, it ensures the secretion of bile acids, which are crucial for lipid emulsification and subsequent lipid absorption in the intestine. Third, it represents an important way for the removal of drugs, toxins, and waste productions from the body.

In humans, hepatic bile is secreted at a rate of 30–40 mL per hour and its volume secreted by the liver is estimated to be between 800 and 1,000 mL per day. Total lipids constitute approximately 3% of hepatic bile by weight, i.e., a total lipid concentration of approximately 3 g/dL. Bile

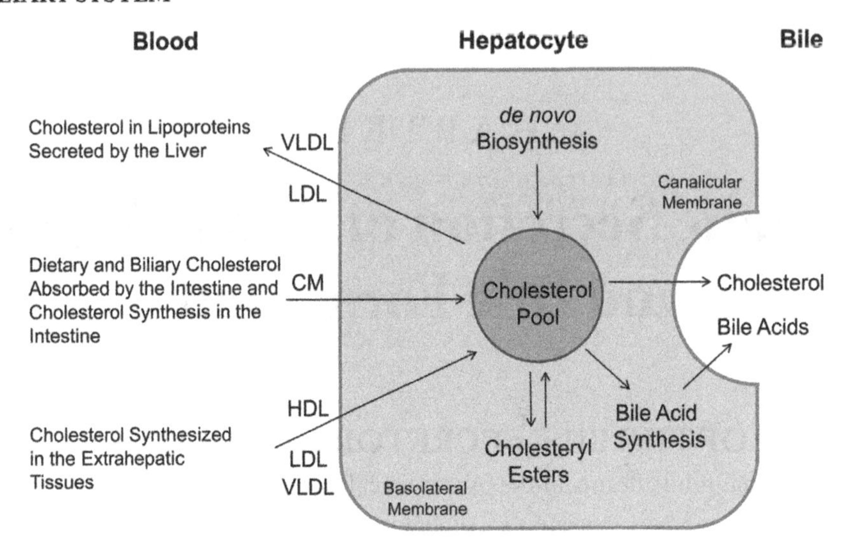

FIGURE 7.1: This diagram shows cholesterol balance across the liver, indicating the major sources for cholesterol entering the hepatocyte and the main pathways for its disposition from the hepatocyte. Abbreviations: CM, chylomicron; HDL, high-density lipoprotein; LDL, low-density lipoprotein; VLDL, very low-density lipoprotein.

acids are the predominant organic components, with concentrations averaging around 20 to 30 mM (12 g/L). Phospholipid and cholesterol concentrations in normal hepatic bile are, on average, 7 mM (5 g/L) and 2 to 3 mM (1 g/L), respectively.

When no dietary cholesterol is consumed, bile contains newly synthesized cholesterol from the liver and preformed cholesterol that reach the liver from several different ways (Figure 7.1). Approximately 80% of biliary total cholesterol is derived from the pools of preformed cholesterol within the liver and about 20% of the cholesterol in bile comes from *de novo* hepatic biosynthesis. The sources of preformed cholesterol are derived from hepatic uptake of plasma lipoproteins such as high-density lipoprotein (HDL), low-density lipoprotein (LDL), and very low-density lipoprotein (VLDL) through their respective receptors on the basolateral membrane of hepatocytes [20]. Consistent with their central role in the reverse cholesterol transport, HDL particles can transfer the cholesterol molecule from the extrahepatic tissues to the liver and are the main lipoprotein source of cholesterol that is targeted for biliary secretion. *De novo* cholesterol biosynthesis in the liver uses acetyl CoA as a substrate and is regulated mainly by the rate-limiting enzyme, 3-hydroxy-3-methylglutaryl coenzyme A (HMG CoA) reductase [142]. This enzyme can be up- or down-regulated depending on the overall cholesterol balance in the liver. An increase in the activity of this rate-limiting enzyme could enhance hepatic secretion of biliary cholesterol [135, 136].

Under conditions of high cholesterol consumption, an appreciable fraction of cholesterol in bile can be derived from the diet through apolipoprotein E-dependent delivery of chylomicron remnants to the liver (Figure 7.1). Dietary cholesterol reaches the liver through the intestinal lymphatic pathways as of chylomicrons, and subsequently, chylomicron remnants after chylomicrons are hydrolyzed by plasma lipoprotein lipase and hepatic lipase [65, 422]. Under the circumstances, newly synthesized cholesterol in the liver is reduced and consists of only approximately 5% of biliary total cholesterol.

The metabolic determinants of the supply of hepatic cholesterol molecules that can be recruited for biliary secretion depend upon the cholesterol input–output balance and its catabolism in the liver (Figure 7.1). Input is dependent on the amount of cholesterol (both unesterified and esterified) taken up by the liver from plasma lipoproteins (LDL > HDL > chylomicron remnants) plus *de novo* hepatic cholesterol biosynthesis. Output is dependent on the amount of cholesterol disposed within the liver after its conversion to cholesteryl ester (to form new VLDL plus ester storage) minus the amount of cholesterol converted to the primary bile acids such as cholic acid (CA) and chenodeoxycholic acid (CDCA). Overall, the liver can systematically regulate the total amount of cholesterol within it, and any excess cholesterol can be handled efficiently.

Although biliary phospholipids are derived from the cell membranes of hepatocytes, their compositions differ significantly. The membranes of hepatocytes contain phosphatidylcholines (lecithins), phosphatidylethanolamines, phosphatidylinositols, phosphatidylserines, and sphingomyelins. The major source of phosphatidylcholine molecules destined for secretion into bile is hepatic synthesis. However, a fraction of biliary phosphatidylcholines may also originate from the phospholipid coat of HDL particles. Approximately 11 g of phospholipids are secreted into bile per day in humans.

More than 95% of the bile acid molecules, which are secreted into bile, have returned to the liver through the enterohepatic circulation by absorption mostly from the distal ileum due to active transport by a specific bile acid transporter, apical sodium-dependent bile acid transporter (ASBT) [229, 262, 423]. As a result, newly synthesized bile acids in the liver contribute only a small fraction (less than 5%) to biliary secretion, which compensate for bile acids that escape intestinal absorption and are lost in the feces. The fecal excretion of bile acids is increased under conditions in which the enterohepatic circulation of bile acids is partially or completely interrupted by surgery, disease states, or drugs (e.g., bile acid-binding resins such as cholestyramine) [352]. A complete interruption of the enterohepatic circulation results in the up-regulation of bile acid synthesis, which could restore bile acid secretion rates to approximately 25% of their normal values [229].

BILIARY LIPID SECRETION

Many animal studies have provided direct evidence showing that bile acids stimulate secretion of vesicles by hepatocytes, and these unilamellar vesicles are always detected in freshly collected hepatic biles [424–429]. When being cultured under specified conditions, rat hepatocytes can form couplets with isolated "bile canaliculi" at interfaces between adjoining cells [430–432]. Using laser

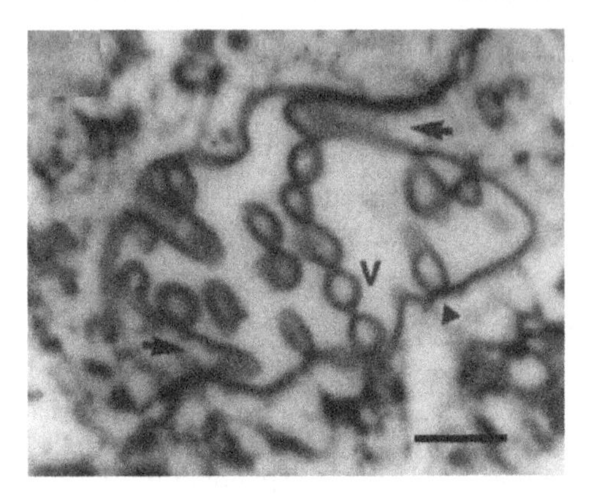

FIGURE 7.2: Electron spectroscopic imaging (right panel) of bile canaliculi from cryofixed rat liver tissues shows distinct vesicles (V) with sharp membrane borders. Vesicles are free within the lumen or attach to the canalicular membrane along linear zones of contact (arrow head). It also shows microvilli closely apposed to the tight junctional region (arrows). Bar = 200 nm. Used with permission from Crawford JM, Möckel GM, Crawford AR, Hagen SJ, Hatch VC, Barnes S, Godleski JJ, Carey MC. *Journal of Lipid Research*. 1995;36(10):2147-2163.

light-scattering techniques, vesicle formation has been observed within these "bile canaliculi" after exposure to bile acids [169, 425, 427, 433, 434]. Moreover, rapid fixation techniques and electron microscopy have provided direct morphologic evidence for vesicle formation at the outer surface of the canalicular membrane (Figure 7.2) [430–432]. It has been proposed that most, if not all, bile acids enter canalicular spaces as monomers and that biliary phospholipids and cholesterol are secreted into bile as unilamellar vesicles [26]. Unilamellar vesicles have been visualized and their sizes have been determined within canalicular spaces of rat liver by transmission electron microscopy and quasi-elastic light-scattering spectroscopy. The radii of these vesicles as determined by both methods generally agree, varying on a range of 60 to 80 nm [435]. Furthermore, to quantify the changing proportions of micelles to vesicles, the bile formation path in a cholesterol-unsaturated fresh hepatic bile of the rat is investigated by dynamic light-scattering spectroscopy [162, 169, 194, 436, 437]. It is found that these vesicles are apparently of more variable size, usually from 40 to 120 nm in radius. The physical–chemical responses of biliary lipids are not instantaneous, and bile that is collected form hepatic ducts as well as from gallbladder is still undergoing equilibration. It has been found that the principal driving forces for biliary lipid secretion are ATP-dependent membrane transporters that are located on the canalicular membrane of the hepatocyte [373, 438–440]. These transport systems most relevant to hepatic secretion of biliary lipids in the human liver are illustrated in Figure 7.3 and listed with their corresponding functions in Table 7.1.

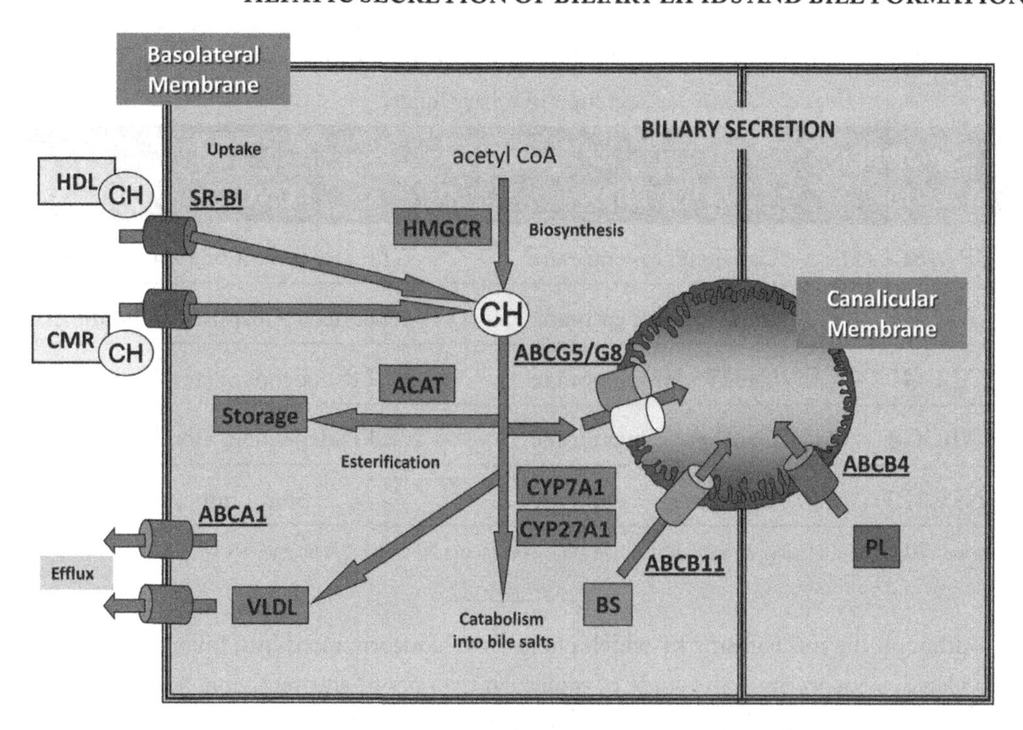

FIGURE 7.3: Uptake, biosynthesis, catabolism, and biliary secretion of cholesterol at the hepatocyte level . The hepatic uptake of cholesterol mediated by scavenger receptor class B type I (SR-BI) for high-density lipoprotein (HDL) and by the chylomicron remnant receptor for chylomicron remnants (CMR). Biosynthesis of hepatic cholesterol is regulated by the rate-limiting enzyme 3-hydroxy-3-methylglutaryl-coenzyme A reductase (HMGCR). Part of the cholesterol is esterified by acyl-coenzyme A:cholesterol acyltransferase (ACAT) for storage in the liver. Some of the cholesterol molecules are used for the formation of VLDL that is secreted into the circulatory system. The ATP-binding cassette (ABC) transporter ABCA1 either directly or indirectly translocates phospholipids and cholesterol to the cell surface, where they appear to form lipid domains that interact with amphipathic α-helices in apolipoproteins. This interaction solubilizes these lipids and generates nascent HDL particles that dissociate from the cell. Binding of apolipoproteins to ABCA1 may also enhance the activity of this lipid-transport pathway. A portion of cholesterol is used for the synthesis of bile salts via the "neutral" and the "acidic" pathways as regulated by two rate-limiting enzymes cholesterol 7α-hydroxylase (CYP7A1) and sterol 27-hydroxylase (CYP27A1), respectively. Hepatic secretion of biliary cholesterol (CH), phospholipids (PL), and bile salts (BS) is determined by three ATP-binding cassette (ABC) transporter proteins, ABCG5/G8, ABCB4, and ABCB11, respectively. Used with permission from Portincasa P, Moschetta A, Di Ciaula A, Pontrelli D, Sasso RC, Wang HH, Wang DQ-H. Pathophysiology and Cholesterol Gallstone Disease in *Biliary Lithiasis: Basic Science, Current Diagnosis and Management*. Editors: Borzellino G and Cordiano C. Springer Italia S.r.l., Milano. 2008. p. 23.

TABLE 7.1: Functions of ATP-binding cassette transporters involved in hepatic secretion of biliary lipids.

TRANSPORTER (GENE)	LOCATION	FUNCTION
BSEP (*ABCB11*)	Canalicular membrane	ATP-dependent bile acid export
ABCB4	Canalicular membrane	ATP-dependent phosphatidylcholine export
ABCG5	Canalicular membrane	ATP-dependent sterol export
ABCG8	Canalicular membrane	ATP-dependent sterol export
NPC1L1	Canalicular membrane	Sterol import

Abbreviations: ABC, ATP-binding cassette; BSEP, bile salt export pump; NPC1L1, Niemann–Pick C1 Like 1.

Although the mechanisms by which cholesterol is incorporated into biliary vesicles are not yet elucidated, evidence from the study of molecular genetics of sitosterolemia has shown that the efflux of biliary cholesterol from the canalicular membrane could be protein-mediated (Figure 7.3) [66, 67, 441–450]. Two plasma membrane proteins, ATP-binding cassette (ABC) transporters ABCG5 and ABCG8, promote cellular efflux of cholesterol, and its significance for bile formation has been examined in genetically modified mice [72, 74, 97–99]. Overexpression of ABCG5/G8 in the liver increases the cholesterol content of gallbladder bile. In contrast, the hepatic secretion rate of biliary cholesterol is reduced in ABCG5/G8 double knockout mice and in ABCG5 or ABCG8 knockout mice. In addition, scavenger receptor class B type I (SR-BI) is localized mainly in sinusoidal, and perhaps, in canalicular membranes of the hepatocyte. In transgenic and knockout mice, biliary secretion of cholesterol varies in proportion to the hepatic expression of SR-BI, and the established contribution of SR-BI to sinusoidal uptake of HDL cholesterol is destined for secretion into bile [451–453]. Furthermore, cholesterol contents of vesicles may be determined potentially by the degree to which cholesterol partitions into phosphatidylcholine-rich microdomains in the canalicular membrane of the hepatocyte. Because of the high affinity of sphingomyelin for cholesterol, microdomains form in membranes with sufficient amounts of cholesterol and sphingomyelin on the canalicular membrane [454–456]. In addition, another possibility is that bile acids could regulate the partitioning of cholesterol into nascent vesicles. Because bile acids reduce the affinity of cholesterol for sphingomyelin, they could induce the migration of cholesterol into phosphatidylcholine-rich microdomains. It is also found that despite a reduced gallstone prevalence rate, the formation of cholesterol gallstones can still be detected in ABCG5/G8 double knockout mice as well as in ABCG5 or ABCG8 knockout mice fed a lithogenic diet. These findings strongly suggest an ABCG5/G8-independent pathway for hepatic secretion of biliary cholesterol and its role in the formation of cholesterol gallstones.

A P-glycoprotein member of the multi-drug resistance gene family, ABCB4 plays an important role in regulating hepatic secretion of biliary phospholipids because the deletion of the *Abcb4* gene results in a complete inhibition on biliary phospholipid secretion in mice [457]. It has been proposed that ABCB4 could be responsible for the translocation or "flip" of phosphatidylcholines from the endoplasmic (inner) to ectoplasmic (outer) leaflet of the canalicular membrane bilayer, and the action of ABCB4 may form phosphatidylcholine-rich microdomains within the outer membrane leaflet [458–462]. Bile acids may partition preferentially into these areas to destabilize the membrane and release phosphatidylcholine-rich vesicles because detergent-like bile acid molecules within the canalicular space could interact with the canalicular membrane. Many observations have suggested that bile acids could promote the vesicular secretion of biliary cholesterol and phosphatidylcholines, although the ectoplasmic leaflet of the canalicular membrane is enriched with cholesterol and sphingomyelin, and is relatively resistant to penetration by bile acids. The biliary secretion of organic anions in laboratory animals does not influence bile acid secretion, but does inhibit the secretion of phospholipids and cholesterol into bile because organic anions can bind bile acids within bile canaliculi and prevent their interactions with the canalicular membrane. Indeed, the mutation of the *ABCB4* gene in humans is the molecular defect underlying type 3 progressive familial intrahepatic cholestasis (PFIC3) [384, 459, 463–466].

Biliary bile acids consist of those that are newly synthesized in the liver and those undergoing enterohepatic cycling [229, 352, 353]. The hepatic secretion of biliary bile acids is induced by ABCB11, a bile acid export pump on the canalicular membrane of hepatocytes [368–372]. Hepatic secretion of bile acids could directly affect cholesterol-phospholipid vesicle secretion [36, 194, 467, 468], although the molecular mechanism by which bile acid secretion is coupled to cholesterol and phospholipid secretion is not known. It has been found that the relationship between bile acid secretion and cholesterol secretion is curvilinear. At low (less than 10 μmol/hr/kg) bile acid secretion rates, more cholesterol is secreted per molecule of bile acid than at higher rates. Although bile acid secretion rates are not usually low in normal subjects, they could diminish during prolonged fasting, during the overnight period, and with substantial bile acid losses such as with a biliary fistula or ileal resection when the liver cannot sufficiently compensate with increased bile acid synthesis. In contrast, at high bile acid secretion rates—for example, during and after eating—biliary saturation is less than during the interprandial period.

In the physiological state, the transport of solutes from the blood to the bile is driven by transport systems in the plasma membrane of the basolateral (sinusoidal) and apical (canalicular) surfaces of hepatocytes (Figure 7.4). Various ATP-binding cassette transporters in the canalicular

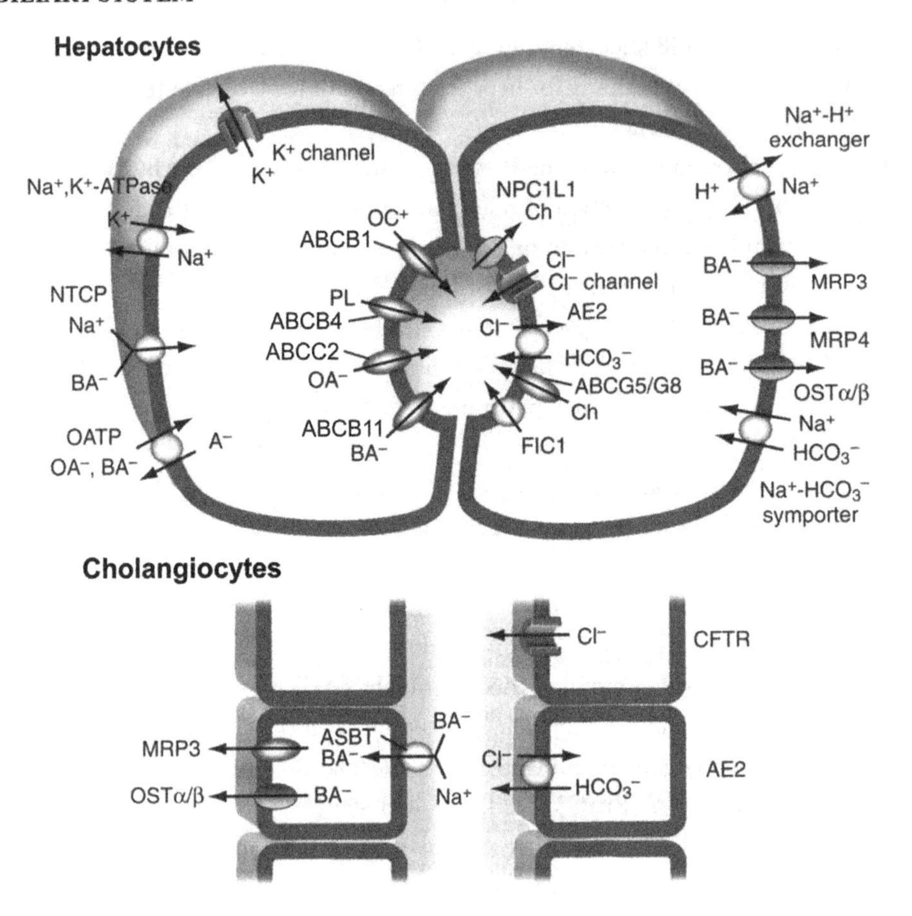

FIGURE 7.4: The hepatocyte and cholangiocyte transporters responsible for the transport of solutes during the bile formation. At the sinusoidal membrane of hepatocytes, there are two systems for bile acid uptake: a sodium-taurocholate cotransporting polypeptide (NTCP) and a sodium-independent organic-anion transporting polypeptide (OATP). Sodium-dependent uptake of bile acids through the NTCP is driven by an inwardly directed sodium gradient generated by Na^+/K^+-ATPase and the membrane potential generated in part by a potassium channel. The sodium-independent bile acid uptake is mediated by the OATP1B1 and OATP1B3. In addition, the sinusoidal membrane contains a sodium–hydrogen exchanger and a sodium-bicarbonate cotransporter (symporter). At the canalicular membrane of hepatocytes, there are several ATP-binding cassette (ABC) transporters. Cholesterol, bile acids, and phospholipids are secreted by ABCG5/G8, ABCB11, and ABCB4, respectively. Sulfated or glucuronidated bile acids are secreted by ABCC2. Drug metabolites are secreted by ABCB1. In addition, the canalicular membrane contains several ATP-independent transport systems, including a chloride channel (distinct from the cystic fibrosis transmembrane regulator protein), a chloride-bicarbonate anion exchanger isoform 2 (AE2) for secretion of bicarbonate, and a glutathione (GSH) transporter. On the apical membrane

membrane mediate the secretion of the main solutes. Bile acids are the most abundant solutes in bile and their transport from plasma into hepatocytes is predominantly mediated by the sodium-taurocholate cotransporter. Bile acids are pumped into primary bile at a concentration of approximately 20 mM. The plasma bile acid concentration is 1000-fold lower. Thus, active transport of solutes across the canalicular membrane of hepatocytes represents the rate-limiting step in bile formation. This unidirectional concentrative step is driven by an array of ATP-dependent export pumps that belong to the ATP-binding cassette family of membrane transporters. It indicates that bile formation is an osmotic secretory process that is driven by the active concentration of bile acids and other biliary constituents in the bile canaliculi. Figure 7.4 depicts the transport systems important in the transport of solutes during the bile formation in the human liver. Table 7.2 lists these transporter proteins with their corresponding functions.

CANALICULAR BILE FLOW

Bile formed by the hepatocytes is secreted in the bile canaliculi and then modified during its passage in the bile ductules and ducts. Bile is an aqueous solution of organic and inorganic compounds. Bile acids, cholesterol, phospholipids, and bile pigments are the major organic compounds. Proteins are present at low concentrations. Metabolites of various endogenous compounds (e.g., hormones) are also found at a trace concentration. Because of the peculiar aggregation properties of the bile acids, which readily form micelles at physiological concentrations, bile is more complex than most other secretions, especially in regard to the osmotic properties of its constituents. The major inorganic electrolytes include Na^+, K^+, Ca^{++}, Mg^{++}, Cl^-, and HCO_3^-, which are present in common duct bile at concentrations closely reflecting those in plasma. Therefore, bile formation involves the secretion of osmotically active inorganic and organic anions into the canalicular lumen, followed by passive water movement. Because of the excellent correlation between bile flow and bile acid output in bile, bile acids are considered to be one of the solutes generating canalicular bile flow. Under several circumstances, however, canalicular bile flow can also be found in the absence of bile acids or at low

---

of cholangiocytes, bile acids are absorbed by the apical sodium-dependent bile acid transporter (ASBT). On the basolateral membrane of cholangiocytes, bile acids may exit by the heteromeric organic solute transporters OSTα/β. Furthermore, cholangiocytes have absorptive functions for a variety of bile solutes, including bile salts, amino acids, and glucose. Cholangiocytes also contain a chloride channel that corresponds to the cystic fibrosis transmembrane regulator (CFTR) and an AE2 for secretion of bicarbonate. Abbreviations: BA bile acids; Ch, cholesterol; OA organic anion; OC organic cation; and PL, phospholipids. Used with permission from Dawson PA. Bile Secretion and the Enterohepatic Circulation in *Sleisenger and Fordtran's Gastrointestinal and Liver Disease.* Editors: Feldman M, Friedman LS, and Brandt L. the 9th Edition. Elsevier Saunders. Philadelphia. 2010; p. 1082.

TABLE 7.2: The hepatocyte plasma membrane bile acid and organic solute transporters involved in bile acid secretion.

NAME	ABBREVIATION (GENE)	LOCATION	FUNCTION
Sodium-taurocholate co-transporter	NTCP (*SLC10A1*)	Basolateral membrane of hepatocytes	Primary carrier for conjugated bile salt uptake from portal blood
Organic anion-transporting polypeptides	OATPs (*SLCO1B1* and *1B3*)	Basolateral membrane of hepatocytes	Broad substrate carriers for sodium-independent uptake of bile salts, organic anions, and other amphipathic organic solutes from portal blood
Organic solute transporter alpha/beta	OSTα/β	Basolateral membrane of hepatocytes, cholangiocites, ileum and proximal tubule of kidney	Heteromeric solute carrier for facilitated transport of bile acids across basolateral membrane of ileum. Expression induced in liver in cholestasis
Organic cation transporter-1	OCT-1 (*SLC22A1*)	Basolateral membrane of hepatocytes	Facilitates sodium-independent hepatic uptake of small organic cations
Organic anion transporter 2	OAT-2 (*SLC22A7*)	Basolateral membrane of hepatocytes	Facilities sodium-independent hepatic uptake of drugs and prostaglandins
Multidrug-resistance-1 P-glycoprotein [a]	MDR1 (*ABCB1*)	Canalicular and cholangiocyte apical membrane	ATP-dependent excretion of various organic cations, xenobiotics, and cytotoxins into bile; barrier function in cholangiocytes

Multidrug-resistance-3 P- glycoprotein (phospholipid transporter) [a]	MDR3 (*ABCB4*)	Canalicular membrane	ATP-dependent translocation of phosphatidylcholine from inner to outer leaflet of membrane bilayer
Bile salt export pump [a]	BSEP (*ABCB11*)	Canalicular membrane	ATP-dependent bile salt transport into bile; stimulate bile salt-dependent bile flow
Multidrug-resistance-associated protein 2 (canalicular multispecific organic anion transporter)[a]	MRP2 (*ABCC2*)	Canalicular membrane	Mediates ATP-dependent multi-specific organic anion transport (e.g., bilirubin diglucuronide) into bile; contributes to bile salt-independent bile flow by GSH transport
Multidrug-resistance-associated protein 3 [a]	MRP3 (*ABCC3*)	Basolateral membrane of hepatocytes and cholangiocytes	Expression induced in cholestasis. Transports bilirubin and bile acid glucuronide conjugates
Multidrug-resistance-associated protein 4 [a]	MRP4 (*ABCC4*)	Basolateral membrane of hepatocytes and apical membrane of proximal tubule of kidney	Expression induced in cholestasis-transports sulfated bile acid conjugates and cyclic nucleotides
Multidrug-resistance-associated protein 6 [a]	MRP6 (*ABCC6*)	Basolateral membrane of hepatocytes	ATP-dependent bile salt transport of organic anions and small peptides. Mutations of MRP6 gene result in pseudoxanthoma elasticum

TABLE 7.2: (*continued*)

NAME	ABBREVIATION (GENE)	LOCATION	FUNCTION
Breast cancer resistance protein [a]	BRCP (*ABCG2*)	Canalicular membrane and proximal tubule of kidney	ATP-dependent multispecific drug transporter, particularly sulfate conjugates; protoporphyrins are endogenous substrate. Substrate overlap with MRP2
Sterolin-1 and -2 [a]	*ABCG5/G8*	Canalicular membrane and apical membrane of intestine	Heteromeric ATP-dependent transport for cholesterol and plant sterols
Multidrug and toxin extrusion protein 1	MATE-1 (*SLC47A1*)	Canalicular membrane and brush border of kidney	Organic cation/H+ exchanger extrudes cationic xenobiotics

[a] These transporters are members of ATP-binding cassette superfamily.

Used with permission from Boyer JL. Adaptive Regulation of Hepatocyte Transporters in Cholestasis in *The Liver: Biology and Pathobiology*. Editors: Arias IM, Alter HJ, Boyer JL, Cohen DE, Fausto N, Shafritz DA, and Wolkoff AW. 5th edition. Wiley-Blackwell, West Sussex, 2009. pp. 683–684.

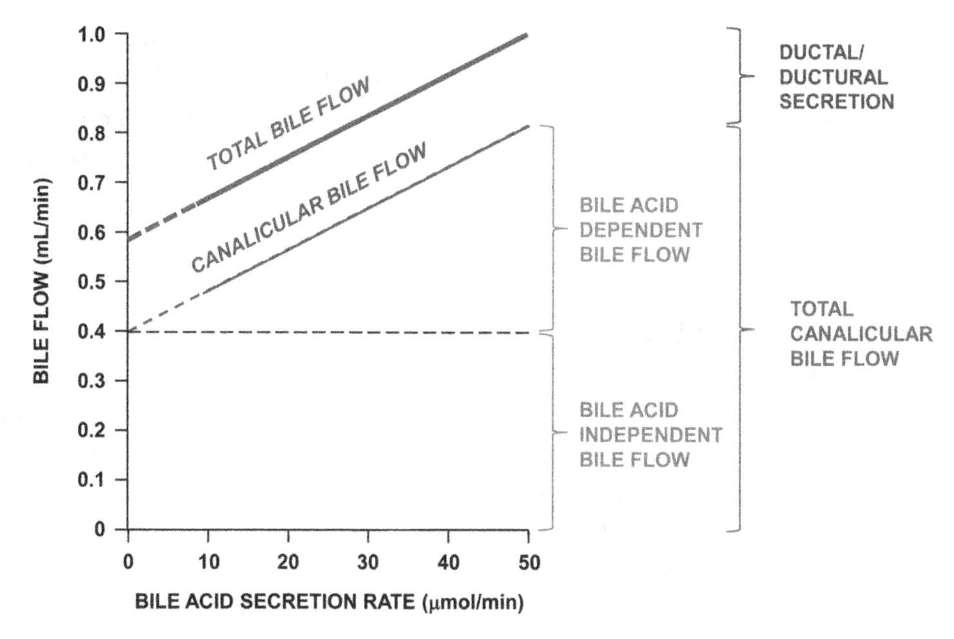

FIGURE 7.5: Schematic representation of the components of bile flow. Total canalicular bile flow (*thin line*) consists of bile acid-dependent and -independent bile flow. Bile acid-dependent bile flow varies directly with hepatic bile acid output. The slope of line is thought to estimate the osmotic activity of the bile acids. When extrapolated to zero bile acid secretion, the line for canalicular bile flow is assumed to estimate the canalicular bile acid-independent secretion. Total bile flow (*thick line*) consists of constant ductal/ductural secretion and total canalicular bile flow. The relation is linear in both total bile flow and total canalicular bile flow.

bile acid outputs [469]. Based on these observations, canalicular bile formation can be divided into two components: bile acid-dependent bile flow (bile flow relating to bile acid secretion) and bile acid-independent flow (bile flow attributed to active secretion of inorganic electrolytes and other solutes) Figure 7.5 shows the relationship between hepatic bile acid secretion rate and bile flow rate. Moreover, total bile flow consists of constant ductal/ductural secretion and total canalicular bile flow. There is significant linear relationship between total bile flow and total canalicular bile flow.

Bile Acid-Dependent Bile Flow

Hepatic ABCB11, the bile acid export pump, can actively secrete bile acids into the canalicular lumen [368–372]. Bile acid concentration in human hepatic bile ranges from 3 to 45 mM. Bile acids are secreted against a concentration gradient. When bile acids are pumped across the canalicular membrane, they stimulate bile production and induce bile flow. An apparently linear relationship between bile acid secretion rates and bile flow has been observed in humans and animal models such as mice, rats, rabbits, dogs, and monkeys [247, 368, 440]. Bile acid-dependent bile flow varies directly with hepatic bile acid output (Figure 7.5). Because canalicular bile flow is linearly related to bile acid secretion rate, the slope of line is thought to estimate the osmotic activity of the bile acids. So, increased bile acid secretion promotes bile flow because bile acids provide an osmotic driving force for filtration or water and electrolytes.

Bile Acid-Independent Bile Flow

As shown in Figure 7.5, total canalicular bile flow contains bile acid-dependent and -independent bile flow. Plots of bile acid secretion rates against bile flow generate a positive intercept upon extrapolation to the flow axis. Thus, when extrapolated to zero bile acid secretion, the line for canalicular bile flow is assumed to estimate the canalicular bile acid-independent secretion. Plots of bile acid secretion rates against bile flow generate a positive intercept upon extrapolation to the flow axis. It is estimated that bile acid-independent fraction of canalicular bile ranges from 1.5 to 2.0 μL/min/kg/body weight in humans. Hepatic secretion of glutathione (GSH) and bicarbonate (HCO_3^-) represents the major components of the bile acid-independent fraction of bile flow. The multidrug resistance-associated protein 2 (MRP2) on the canalicular membrane of hepatocyte plays a crucial role in hepatic GSH secretion. Intraluminal catabolism of GSH by γ-glutamyl transpeptidase (GGTP) also contributes to the osmotic driving force for canalicular bile formation. Because of the activity of the membrane-bound enzyme Na^+,K^+-ATPase, active sodium transport into the canaliculi induces bile acid-independent bile flow. In the isolated perfused rat liver, perfusion with a bicarbonate-free solution reduces the bile acid-independent flow by 50%. Besides the ATP-dependent secretion of organic anions into bile, hepatic and biliary ATP-independent secretion of HCO_3^- via the chloride-bicarbonate anion exchanger isoform 2 (AE2) induces bile acid-independent bile flow. The majority of this HCO_3^- secretion occurs at the level of the bile duct epithelial cells in response to stimulation by a variety of hormones and neuropeptides, such as secretion and vasoactive intestinal peptide.

PHYSICAL STATES OF BILIARY LIPIDS

Because bile is an aqueous solution and cholesterol is virtually insoluble in water, the mechanisms for the solubilization of cholesterol in bile are complex. Lecithins are also insoluble in water. So, both cholesterol and lecithins have to travel together with bile acids for transport in bile. Both micelles and vesicles are two main types of macromolecular aggregates in bile (Figure 7.6). It has been found that bile acids, either as simple solutions or as mixtures together with cholesterol and lecithin, can form micellar solutions [26, 102–104, 189, 470–475]. A micelle is a colloidal aggregation of molecules of an amphipathic compound (i.e., bile acids) in which the hydrophobic portion of each molecule faces inward and the hydrophilic groups point outward. The cholesterol molecule can be solubilized possibly within the hydrophobic center of the micelle. Furthermore, biliary vesicles can also act as a cholesterol solubilizer in bile, which are spherical membrane bilayers that contain mainly lecithin and cholesterol with only traces of bile acids. Vesicles are unilamellar (i.e., a single bilayer that encircles an aqueous core) or multilamellar (i.e., contain multiple concentric spherical bilayers) [474, 475]. The precise compositions and proportions of micelles and vesicles are determined essentially by the concentrations of biliary lipids, which vary considerably within the biliary tree and gallbladder. In dilute hepatic bile (i.e., a total lipid concentration less than 3 g/dL), vesicles are extremely stable and do not aggregate, fuse, or nucleate cholesterol crystals. Despite solid cholesterol monohydrate crystals are one of the equilibrium phases, they form very slowly in dilute hepatic bile. However, when bile is concentrated by the gallbladder and total lipid concentration is increased and reaches at about ~10 g/dL, vesicle instability is significantly increased and cholesterol precipitation is greatly accelerated.

BILIARY MICELLES

Because bile acids have both hydrophilic and hydrophobic areas with the property of amphiphilicity [149], they are soluble in aqueous solutions to varying degrees, depending on the number and characteristics of hydroxyl groups and side chains, as well as on the composition of the particular aqueous solution [150, 423]. Bile acid monomers can aggregate spontaneously to form simple micelles when a critical micellar concentration (CMC) is exceeded [154, 156, 158]. These simple micelles (~3 nm in diameter) are small, thermodynamically stable aggregates that can solubilize cholesterol. They are shaped like disks, which has been characterized in physiologically relevant model biles and in native human and animal biles. The formation of simple micelles of bile acids depends mainly on their concentrations. Thus, micelles form at, but not below, a CMC of bile acids in bile, which is about 2 mmol/L. Also, simple micelles are capable of solubilizing and incorporating phospholipids, which are referred to as mixed micelles. These alterations enable mixed micelles to solubilize at least three times the amount of cholesterol solubilized by simple micelles. Mixed micelles (~4-8 nm in diameter) are large, thermodynamically stable aggregates that are composed of bile acids, cholesterol, and phospholipids (Figure 7.6). Their sizes vary depending on the relative proportion of bile acids and phospholipids. The shape of a mixed micelle is that of a lipid bilayer with the hydrophilic groups of the bile acids and phospholipids aligned on the "outside" of the bilayer and the hydrophobic groups on the "inside." Therefore, cholesterol molecules can be solubilized on the inside of the bilayer away from the aqueous areas on the outside. The amount of cholesterol that can be solubilized depends on the relative proportions of bile acids, and maximal solubility occurs

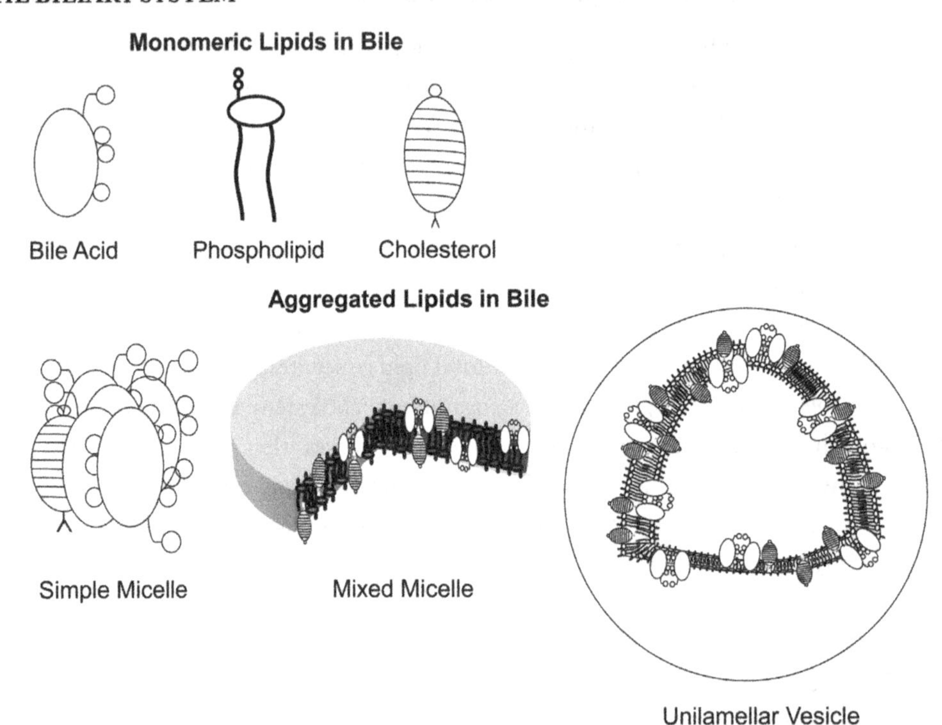

FIGURE 7.6: Physical states of lipids in human bile (not drawn to scale). Cholesterol and phospholipids are highly insoluble in water. In contrast, bile acids are highly water soluble. Bile acids remain monomeric up to their critical micellar concentration (average 1 to 3 mM). Excess bile acids at greater than this concentration can self-aggregate as micelles. Simple bile acid micelles (~1 to 2 nm in diameter) can bind a molecule of cholesterol, which increases the aqueous solubility of cholesterol. With typical gallbladder lipid compositions, simple bile acid and mixed bile acid-lecithin micelles co-exist in a ratio of 1:5. The site of attachment for cholesterol molecules on simple micelles is on the exterior hydrophilic surface. In mixed micelles, cholesterol is solubilized within the micelles. Phospholipids in an aqueous environment self-aggregate to form stable bilayers. Under the circumstances, a large amount of the cholesterol molecules is inserted into these bilayers between the hydrophobic acyl chains of the phospholipids. The ratio of unilamellar vesicles to micelles depends on the bile acid and cholesterol concentrations of the bile, which is greatest in bile with low bile acid and high cholesterol concentrations. Furthermore, at low bile acid concentrations, these biliary lipids often form large unilamellar (~40 to 100 nm in diameter) or multilamellar (~300 to 500 nm in diameter) layers of vesicles. High concentrations of bile acids can dissolve these vesicles to form small mixed micelles (~4 to 8 nm in diameter).

when the molar ratio of phospholipids to bile acids is between 0.2 and 0.3. Furthermore, solubility of cholesterol in mixed micelles is enhanced when the concentration of total lipids (bile acids, phospholipids, and cholesterol) in bile is increased.

BILIARY VESICLES, NONMICELLAR CARRIERS OF CHOLESTEROL IN BILE

Human studies have found that some normal subjects without solid plate-like cholesterol monohydrate crystals or cholesterol gallstones have gallbladder bile that is supersaturated with cholesterol. In addition, there is a difference in metastability between patients with cholesterol gallstones and subjects without gallstones [103, 117, 471, 474–476]. It has also been found that some patients with cholesterol gallstones have cholesterol solubility greatly exceeding the limits of metastability. These findings strongly suggest that besides micellar systems of bile acids, other mechanisms must exist for the solubilization of cholesterol in bile. By using techniques such as quasi-elastic light-scattering spectroscopy and electron microscopy to examine native and model biles, a more complex mechanism on cholesterol solubilization in bile has been defined. It has been found that gallbladder bile from many healthy individuals are supersaturated with cholesterol, indicating that cholesterol concentrations has exceeded what can be solubilized by micellar particles. Biliary vesicles were first reported in 1983 [429, 434], and these findings clarified substantially our understanding of cholesterol solubility in bile, including metastability, liquid crystal formation, and cholesterol nucleation and crystallization in bile. Vesicles are unilamellar spherical structures, and contain phospholipids, cholesterol, and little, if any, bile acids. Thus, vesicles (~40 to 100 nm in diameter) are substantially larger than both simple and mixed micelles, but much smaller than liquid crystals (~500 nm in diameter) that are composed of multilamellar spherical structures (Figure 7.2). Because vesicles are present in large quantities in hepatic bile, they are presumably secreted by the hepatocyte. As a result of the time-dependent physical-chemistry of postcanalicular events during bile formation and because of very slow disappearance of vesicles by micellar dissolution, unilamellar vesicles are often detected in all freshly collected cholesterol-unsaturated bile and are physically indistinguishable from those in supersaturated bile. Of note is that cholesterol crystals and gallstones never form in hepatic bile although dilute hepatic bile is always supersaturated with cholesterol. Thus, the vesicle is a particle responsible for solubilizing biliary cholesterol in excess of what could be solubilized in mixed micelles. Also, it has been proven that these cholesterol-rich vesicles are very stable in dilute hepatic bile. However, when cholesterol concentrations are significantly increased in bile, the unilamellar vesicles can aggregate and fuse to form large multilamellar vesicles (also known as liposomes or liquid crystals). It has been found that solid plate-like cholesterol monohydrate crystals could nucleate from multilamellar vesicles in concentrated gallbladder bile.

CO-EXISTENCE AND INTERCONVERSION OF MICELLES AND VESICLES IN BILE

Although vesicles are relatively static structures, the dynamics of vesicles, in terms of size, composition, and interaction with each other and with micelles, is influenced by several factors such as the total lipid concentration of bile and the relative ratio of cholesterol, phospholipids, and bile acids in bile. The relative ratio of these three lipids is influenced by the fasting or the feeding states through alterations in hepatic secretion rates of biliary lipids. For example, during the fasting period, biliary bile acid output is relatively low. Consequently, the ratio of cholesterol to bile acids is increased and cholesterol is carried more in vesicles than in micelles. In contrast, during meals, biliary bile acid output is higher and more cholesterol is solubilized in micelles. In addition, when the concentration of bile acids is relatively low, especially in dilute hepatic bile, vesicles are relatively stable, and only some vesicles could be converted to micelles. In contrast, vesicles could be transformed or converted completely to mixed micelles with increasing bile acid concentration in concentrated gallbladder bile. Because relatively more phospholipids than cholesterol can be transferred from vesicles to mixed micelles, the residual vesicles are remodeled, which may be enriched in cholesterol relative to phospholipids. If the remaining vesicles have a relatively low cholesterol/phospholipid ratio (less than 1), they are relatively stable. However, if the cholesterol/phospholipid ratio in vesicles is greater than 1, vesicles become increasingly unstable. These cholesterol-rich vesicles may transfer some cholesterol to less cholesterol-rich vesicles or to micelles, or may fuse or aggregate to form much larger (~500 nm in diameter) multilamellar vesicles (i.e., liposomes or liquid crystals). These liquid crystals are visible through polarizing light microscopy as lipid circular droplets with typical birefringence in the shape of a Maltese cross. Liquid crystals are inherently unstable and could form solid plate-like cholesterol monohydrate crystals, which is termed cholesterol nucleation. As a result, the nucleation of these solid cholesterol monohydrate crystals could result in a marked decrease in the amount of cholesterol contained in vesicles but not of cholesterol in micelles, supporting the concept that vesicles could serve as the primary source of cholesterol for nucleation and crystallization.

Although bile is rarely at equilibrium, the physical states of biliary lipids at equilibrium have provided the physical–chemical basis for describing biliary lipid aggregation in hepatic and gallbladder biles [104, 105, 189]. Because bile is gradually concentrated within the biliary tree, flowing from the liver to the gallbladder, bile acid concentrations approach their CMC values. When this occurs, bile acids begin to modify the structure of phospholipid-rich vesicles that are secreted into bile by the hepatocyte. These interactions indicate the start of a complex series of molecular rearrangements of these biliary lipids that ultimately form simple and mixed micelles. In human bile supersaturated with cholesterol, there are two pathways for forming cholesterol-rich vesicles from phospholipid-rich vesicles that are assembled at the canalicular membrane of hepatocytes. Because bile acids solubilize phospholipids more efficiently than cholesterol, cholesterol-rich vesicles may form when bile acids preferentially extract phospholipid molecules directly from phospholipid-rich vesicles [120, 474, 475]. The second possibility that has been observed in model bile systems is rapid dissolution of phospholipid-rich vesicles by bile acids. This alteration produces unstable mixed micelles with excess cholesterol. Consequently, structural rearrangements of these unstable micellar particles could result in the formation of cholesterol-rich vesicles.

• • • •

Summary

The exponential expansion of knowledge in the field of hepatobiliary diseases makes systematic revisions of current concepts almost mandatory, nowadays. Advances are made every day in virtually all fields of science, including genetics, molecular and cellular biology, diagnosis and treatment. As science progresses, technology is also continuously evolving: this eBook is an example of integrated efforts to construct a manageable text, easy-to-study, updated and critical, by using the Web as the main source of knowledge.

The biliary system is comprised of a fantastic network of microscopic and macroscopic structures accomplishing complex tasks, i.e., the formation of bile, an aqueous fluid in which considerable amount of otherwise immiscible cholesterol, is transported by other lipids such as bile acids and phospholipids. Overall, this task is achieved via multiple pathways and precise steps involving the absorption, transport, secretion, re-absorption, and re-secretion of lipids, water, ions, and cations across the liver and biliary tract. Tens of different transporters are continuously regulated—depending on local and general stimuli in health and disease—to play a precise hierarchical "concert" of solutes and water flowing from blood, to liver, to bile ducts, to intestine, and back to liver.

By guiding the readers through the various aspects of anatomy, physiology, and biochemistry of all "players" involved in bile formation, this eBook becomes a *compendium* of recent progresses in understanding the molecular mechanisms of cholesterol and bile acid metabolism. Biliary lipid metabolism and regulation by nuclear receptors is also examined in the hepatobiliary system.

When mechanisms governing the hepatic secretion of biliary lipids fail, the impact on major systemic diseases becomes deleterious. Examples in this respect are atherosclerosis, liver steatosis, cholesterol cholelitiasis, cholestasis, and liver cirrhosis. This eBook depicts therefore the backbone structure of mainly physiological events, which need to be known and preserved to avoid such abnormalities.

Also, updated references, tables, and figures enrich this "virtual" textbook.

Lastly, without prior fruitful interaction with myriad of medical students, clinical and research fellows, residents, and biomedical scientists, it would have been impossible to assemble this work. To all, as Editors, we feel indebted.

Acknowledgments

We are greatly indebted to Helen H. Wang, Ornella de Bari, Tony Y. Wang, and Rajshree Chaudhari for superb word-processing assistance, as well as for preparing figures and tables. This work was supported in part by research grants DK54012 and DK73917 (D.Q.-H.W.) from the National Institutes of Health (US Public Health Service), ORBA09XZZT and ORBA10ROPA (P.P.) from University of Bari, and FIRB 2003 RBAU01RANB002 (P.P.) from the Italian Ministry of University and Research. P.P. was a recipient of the short-term mobility grant 2005 from the Italian National Research Council (CNR).

References

[1] Borley NR: Liver. In: Standring S, ed. Gray's Anatomy: The Anatomical Basis of Clinical Practice. 39 ed. London: Elsevier Churchill Livingstone, 2005; pp. 1213–25.

[2] Deshpande RR, Heaton ND, Rela M. Surgical anatomy of segmental liver transplantation. *Br J Surg* 2002;89: pp. 1078–88.

[3] Healey JE, Jr., Schroy PC. Anatomy of the biliary ducts within the human liver; analysis of the prevailing pattern of branchings and the major variations of the biliary ducts. *AMA Arch Surg* 1953;66: pp. 599–616.

[4] Skandalakis JE, Skandalakis LJ, Skandalakis PN, Mirilas P. Hepatic surgical anatomy. *Surg Clin North Am* 2004;84: pp. 413–35, viii.

[5] Misdraji J: Embryology, Anatomy, Histology, and Developmental Anomalies of the Liver. In: Feldman M, Friedman LS, Brandt L, eds. Sleisenger and Fordtran's Gastrointestinal and Liver Disease. 9 ed. Philadelphia: Elsevier Saunders, 2010; pp. 1201–6.

[6] Borley NR: Gallbladder and Biliary Tree. In: Standring S, ed. Gray's Anatomy: The Anatomical Basis of Clinical Practice. 39 ed. London: Elsevier Churchill Livingstone, 2005; pp. 1227–30.

[7] Couinaud C. Le Foie: Studes Anatomiques et Chirurgicales. Paris: Masson & Cie, 1957; pp. 9–12.

[8] Strasberg SM. Terminology of liver anatomy and liver resections: coming to grips with hepatic Babel. *J Am Coll Surg* 1997;184: pp. 413–34.

[9] Takasaki S, Hano H. Three-dimensional observations of the human hepatic artery (Arterial system in the liver). *J Hepatol* 2001;34:4 pp. 55–66.

[10] Lamah M, Karanjia ND, Dickson GH. Anatomical variations of the extrahepatic biliary tree: review of the world literature. *Clin Anat* 2001;14: pp. 167–72.

[11] Adkins RB, Jr., Chapman WC, Reddy VS. Embryology, anatomy, and surgical applications of the extrahepatic biliary system. *Surg Clin North Am* 2000;80: pp. 363–79.

[12] Frierson HF, Jr. The gross anatomy and histology of the gallbladder, extrahepatic bile ducts, Vaterian system, and minor papilla. *Am J Surg Pathol* 1989;13: pp. 146–62.

[13] Jones AL, Schmucker DL, Renston RH, Murakami T. The architecture of bile secretion. A morphological perspective of physiology. *Dig Dis Sci* 1980;25: pp. 609–29.

[14] Roskams TA, Theise ND, Balabaud C, Bhagat G, Bhathal PS, Bioulac-Sage P, Brunt EM, et al. Nomenclature of the finer branches of the biliary tree: canals, ductules, and ductular reactions in human livers. *Hepatology* 2004;39: pp. 1739–45.

[15] Ludwig J, Ritman EL, LaRusso NF, Sheedy PF, Zumpe G. Anatomy of the human biliary system studied by quantitative computer-aided three-dimensional imaging techniques. *Hepatology* 1998;27: pp. 893–9.

[16] Avisse C, Flament JB, Delattre JF. Ampulla of Vater. Anatomic, embryologic, and surgical aspects. *Surg Clin North Am* 2000;80: pp. 201–12.

[17] Suchy FJ: Anatomy, Histology, Embryology, Developmental Anomalies, and Pediatric Disorders of the Biliary Tract. In: Feldman M, Friedman LS, Brandt L, eds. Sleisenger and Fordtran's Gastrointestinal and Liver Disease. 9 ed. Philadelphia: Elsevier Saunders, 2010; pp. 1045–66.

[18] Meilstrup JW, Hopper KD, Thieme GA. Imaging of gallbladder variants. *AJR Am J Roentgenol* 1991;157: pp. 1205–8.

[19] Suzuki M, Akaishi S, Rikiyama T, Naitoh T, Rahman MM, Matsuno S. Laparoscopic cholecystectomy, Calot's triangle, and variations in cystic arterial supply. *Surg Endosc* 2000;14: pp. 141–4.

[20] Cohen DE: Lipoprotein Metabolism and Cholesterol Balance. In: Arias IM, Alter HJ, Boyer JL, Cohen DE, Fausto N, Shafritz DA, Wolkoff AW, eds. The Liver: Biology and Pathobiology. 5 ed. West Sussex: Wiley-Blackwell, 2009; pp. 271–85.

[21] Brown MS, Goldstein JL. The receptor model for transport of cholesterol in plasma. *Ann N Y Acad Sci* 1985;454: pp. 178–82.

[22] Brown MS, Goldstein JL. Lipoprotein receptors in the liver. Control signals for plasma cholesterol traffic. *J Clin Invest* 1983;72: pp. 743–7.

[23] Brown MS, Goldstein JL. Multivalent feedback regulation of HMG CoA reductase, a control mechanism coordinating isoprenoid synthesis and cell growth. *J Lipid Res* 1980;21: pp. 505–17.

[24] Brown MS, Goldstein JL. Receptor-mediated endocytosis: insights from the lipoprotein receptor system. *Proc Natl Acad Sci U S A* 1979;76: pp. 3330–7.

[25] Brown MS, Kovanen PT, Goldstein JL. Regulation of plasma cholesterol by lipoprotein receptors. *Science* 1981;212: pp. 628–35.

[26] Wang DQ, Cohen DE, Carey MC. Biliary lipids and cholesterol gallstone disease. *J Lipid Res* 2009;50 Suppl: pp. S406–11.

[27] Wang HH, Portincasa P, Wang DQ. Molecular pathophysiology and physical chemistry of cholesterol gallstones. *Front Biosci* 2008;13: pp. 401–23.

[28] Carey MC, Lamont JT. Cholesterol gallstone formation. 1. Physical-chemistry of bile and biliary lipid secretion. *Prog Liver Dis* 1992;10: pp. 139–63.

[29] Spady DK, Woollett LA, Dietschy JM. Regulation of plasma LDL-cholesterol levels by dietary cholesterol and fatty acids. *Annu Rev Nutr* 1993;13: pp. 355–81.

[30] Dietschy JM. Regulation of cholesterol metabolism in man and in other species. *Klin Wochenschr* 1984;62: pp. 338–45.

[31] Grundy SM. Cholesterol metabolism in man. *West J Med* 1978;128: pp. 13–25.

[32] Goldstein JL, Brown MS. Lipoprotein receptors and the control of plasma LDL cholesterol levels. *Eur Heart J* 1992;13 Suppl B: pp. 34–6.

[33] Goldstein JL, Brown MS. Progress in understanding the LDL receptor and HMG-CoA reductase, two membrane proteins that regulate the plasma cholesterol. *J Lipid Res* 1984;25: pp. 1450–61.

[34] Goldstein JL, Brown MS. Familial hypercholesterolemia: pathogenesis of a receptor disease. *Johns Hopkins Med J* 1978;143: pp. 8–16.

[35] Goldstein JL, Brown MS. Lipoprotein receptors, cholesterol metabolism, and atherosclerosis. *Arch Pathol* 1975;99: pp. 181–4.

[36] Carey MC, Cohen DE. Update on physical state of bile. *Ital J Gastroenterol* 1995;27: pp. 92–100.

[37] Freeman JB, Meyer PD, Printen KJ, Mason EE, DenBesten L. Analysis of gallbladder bile in morbid obesity. *Am J Surg* 1975;129: pp. 163–6.

[38] Shaffer EA, Braasch JW, Small DM. Bile composition at and after surgery in normal persons and patients with gallstones. Influence of cholecystectomy. *N Engl J Med* 1972;287: pp. 1317–22.

[39] Shaffer EA, Small DM. Biliary lipid secretion in cholesterol gallstone disease. The effect of cholecystectomy and obesity. *J Clin Invest* 1977;59: pp. 828–40.

[40] Shaffer EA, Small DM. Gallstone disease: pathogenesis and management. *Curr Probl Surg* 1976;13: pp. 3–72.

[41] Donovan JM, Carey MC. Separation and quantitation of cholesterol "carriers" in bile. *Hepatology* 1990;12:94S–104S; discussion 104S–5S.

[42] Small DM. Cholesterol nucleation and growth in gallstone formation. *N Engl J Med* 1980;302: pp. 1305–7.

[43] Bourges M, Small DM, Dervichian DG. Biophysics of lipid associations. 3. The quaternary systems lecithin-bile salt-cholesterol-water. *Biochim Biophys Acta* 1967;144: pp. 189–201.

[44] Bourges M, Small DM, Dervichian DG. Biophysics of lipidic associations. II. The ternary systems: cholesterol-lecithin-water. *Biochim Biophys Acta* 1967;137: pp. 157–67.

[45] Brecher P, Chobanian J, Small DM, Chobanian AV. The use of phospholipid vesicles for in vitro studies on cholesteryl ester hydrolysis. *J Lipid Res* 1976;17: pp. 239–47.

[46] Carey MC. Aqueous bile salt-lecithin-cholesterol systems: equilibrium aspects. *Hepatology* 1984;4: pp. 151S–4S.

[47] Gantz DL, Wang DQ, Carey MC, Small DM. Cryoelectron microscopy of a nucleating model bile in vitreous ice: formation of primordial vesicles. *Biophys J* 1999;76: pp. 1436–51.

[48] Hay DW, Cahalane MJ, Timofeyeva N, Carey MC. Molecular species of lecithins in human gallbladder bile. *J Lipid Res* 1993;34: pp. 759–68.

[49] Andersen JM, Dietschy JM. Regulation of sterol synthesis in 16 tissues of rat. I. Effect of diurnal light cycling, fasting, stress, manipulation of enterohepatic circulation, and administration of chylomicrons and triton. *J Biol Chem* 1977;252: pp. 3646–51.

[50] Andersen JM, Dietschy JM. Regulation of sterol synthesis in 15 tissues of rat. II. Role of rat and human high and low density plasma lipoproteins and of rat chylomicron remnants. *J Biol Chem* 1977;252: pp. 3652–9.

[51] Spady DK, Dietschy JM. Rates of cholesterol synthesis and low-density lipoprotein uptake in the adrenal glands of the rat, hamster and rabbit in vivo. *Biochim Biophys Acta* 1985;836: pp. 167–75.

[52] Turley SD, Dietschy JM. The Metabolism and Excretion of Cholesterol by the Liver. In: Arias IM, Jakoby WB, Popper H, D. S, A. SD, eds. The Liver: Biology and Pathobiology. 2 ed. New York: Raven Press, 1988; pp. 617–41.

[53] Dietschy JM, Turley SD. Control of cholesterol turnover in the mouse. *J Biol Chem* 2002; 277: pp. 3801–4.

[54] Spady DK, Turley SD, Dietschy JM. Rates of low density lipoprotein uptake and cholesterol synthesis are regulated independently in the liver. *J Lipid Res* 1985;26: pp. 465–72.

[55] Spady DK, Turley SD, Dietschy JM. Receptor-independent low density lipoprotein transport in the rat in vivo. Quantitation, characterization, and metabolic consequences. *J Clin Invest* 1985;76: pp. 1113–22.

[56] Turley SD, Dietschy JM. Sterol absorption by the small intestine. *Curr Opin Lipidol* 2003;14: pp. 233–40.

[57] Wang DQ, Cohen DE: Absorption and Excretion of Cholesterol and Other Sterols. In: Ballantyne CM, ed. Lipidology in the Treatment and Prevention of Cardiovascular Disease (Clinical Lipidology: A Companion to Braunwald's Heart Disease). 1 ed. Philadelphia: Elsevier Saunders, 2008; pp. 26–44.

[58] Tso P, Fujimoto K. The absorption and transport of lipids by the small intestine. *Brain Res Bull* 1991;27: pp. 477–82.

[59] Wang DQ, Lee SP. Physical chemistry of intestinal absorption of biliary cholesterol in mice. *Hepatology* 2008;48: pp. 177–85.

[60] Wang DQ, Paigen B, Carey MC. Genetic factors at the enterocyte level account for variations in intestinal cholesterol absorption efficiency among inbred strains of mice. *J Lipid Res* 2001;42: pp. 1820–30.

[61] Bhattacharyya AK, Eggen DA. Relationships between dietary cholesterol, cholesterol absorption, cholesterol synthesis, and plasma cholesterol in rhesus monkeys. *Atherosclerosis* 1987;67: pp. 33–9.

[62] Trautwein EA, Forgbert K, Rieckhoff D, Erbersdobler HF. Impact of beta-cyclodextrin and resistant starch on bile acid metabolism and fecal steroid excretion in regard to their hypolipidemic action in hamsters. *Biochim Biophys Acta* 1999;1437: pp. 1–12.

[63] Turley SD, Daggy BP, Dietschy JM. Effect of feeding psyllium and cholestyramine in combination on low density lipoprotein metabolism and fecal bile acid excretion in hamsters with dietary-induced hypercholesterolemia. *J Cardiovasc Pharmacol* 1996;27: pp. 71–9.

[64] Turley SD, Daggy BP, Dietschy JM. Cholesterol-lowering action of psyllium mucilloid in the hamster: sites and possible mechanisms of action. *Metabolism* 1991;40: pp. 1063–73.

[65] Wang DQ. Regulation of intestinal cholesterol absorption. *Annu Rev Physiol* 2007;69: pp. 221–48.

[66] Berge KE, Tian H, Graf GA, Yu L, Grishin NV, Schultz J, Kwiterovich P, et al. Accumulation of dietary cholesterol in sitosterolemia caused by mutations in adjacent ABC transporters. *Science* 2000;290: pp. 1771–5.

[67] Berge KE, von Bergmann K, Lutjohann D, Guerra R, Grundy SM, Hobbs HH, Cohen JC. Heritability of plasma noncholesterol sterols and relationship to DNA sequence polymorphism in ABCG5 and ABCG8. *J Lipid Res* 2002;43: pp. 486–94.

[68] Graf GA, Cohen JC, Hobbs HH. Missense mutations in ABCG5 and ABCG8 disrupt heterodimerization and trafficking. *J Biol Chem* 2004;279: pp. 24881–8.

[69] Graf GA, Li WP, Gerard RD, Gelissen I, White A, Cohen JC, Hobbs HH. Coexpression of ATP-binding cassette proteins ABCG5 and ABCG8 permits their transport to the apical surface. *J Clin Invest* 2002;110: pp. 659–69.

[70] Graf GA, Yu L, Li WP, Gerard R, Tuma PL, Cohen JC, Hobbs HH. ABCG5 and ABCG8 are obligate heterodimers for protein trafficking and biliary cholesterol excretion. *J Biol Chem* 2003;278: pp. 48275–82.

[71] Hubacek JA, Berge KE, Cohen JC, Hobbs HH. Mutations in ATP-cassette binding proteins G5 (ABCG5) and G8 (ABCG8) causing sitosterolemia. *Hum Mutat* 2001;18: pp. 359–60.

[72] Yu L, Li-Hawkins J, Hammer RE, Berge KE, Horton JD, Cohen JC, Hobbs HH. Overexpression of ABCG5 and ABCG8 promotes biliary cholesterol secretion and reduces fractional absorption of dietary cholesterol. *J Clin Invest* 2002;110: pp. 671–80.

[73] Lee MH, Lu K, Hazard S, Yu H, Shulenin S, Hidaka H, Kojima H, et al. Identification of a gene, ABCG5, important in the regulation of dietary cholesterol absorption. *Nat Genet* 2001;27: pp. 79–83.

[74] Wang HH, Patel SB, Carey MC, Wang DQ. Quantifying anomalous intestinal sterol uptake, lymphatic transport, and biliary secretion in Abcg8(-/-) mice. *Hepatology* 2007;45: pp. 998–1006.

[75] Duan LP, Wang HH, Ohashi A, Wang DQ. Role of intestinal sterol transporters Abcg5, Abcg8, and Npc1l1 in cholesterol absorption in mice: gender and age effects. *Am J Physiol Gastrointest Liver Physiol* 2006;290: pp. G269–76.

[76] Duan LP, Wang HH, Wang DQ. Cholesterol absorption is mainly regulated by the jejunal and ileal ATP-binding cassette sterol efflux transporters Abcg5 and Abcg8 in mice. *J Lipid Res* 2004;45: pp. 1312–23.

[77] Altmann SW, Davis HR, Jr., Zhu LJ, Yao X, Hoos LM, Tetzloff G, Iyer SP, et al. Niemann–Pick C1 Like 1 protein is critical for intestinal cholesterol absorption. *Science* 2004;303: pp. 1201–4.

[78] Davis HR, Jr., Altmann SW. Niemann–Pick C1 Like 1 (NPC1L1) an intestinal sterol transporter. *Biochim Biophys Acta* 2009;1791: pp. 679–83.

[79] Davis HR, Jr., Basso F, Hoos LM, Tetzloff G, Lally SM, Altmann SW. Cholesterol homeostasis by the intestine: lessons from Niemann–Pick C1 Like 1 [NPC1L1). *Atheroscler Suppl* 2008;9: pp. 77–81.

[80] Davis HR, Jr., Zhu LJ, Hoos LM, Tetzloff G, Maguire M, Liu J, Yao X, et al. Niemann–Pick C1 Like 1 (NPC1L1) is the intestinal phytosterol and cholesterol transporter and a key modulator of whole-body cholesterol homeostasis. *J Biol Chem* 2004;279: pp. 33586–92.

[81] Garcia-Calvo M, Lisnock J, Bull HG, Hawes BE, Burnett DA, Braun MP, Crona JH, et al. The target of ezetimibe is Niemann–Pick C1-Like 1 (NPC1L1). *Proc Natl Acad Sci U S A* 2005;102: pp. 8132–7.

[82] Lammert F, Wang DQ. New insights into the genetic regulation of intestinal cholesterol absorption. *Gastroenterology* 2005;129: pp. 718–34.

[83] Brown MS, Brannan PG, Bohmfalk HA, Brunschede GY, Dana SE, Helgeson J, Goldstein JL. Use of mutant fibroblasts in the analysis of the regulation of cholesterol metabolism in human cells. *J Cell Physiol* 1975;85: pp. 425–36.

[84] Brown MS, Dana SE, Dietschy JM, Siperstein MD. 3-Hydroxy-3-methylglutaryl coenzyme A reductase. Solubilization and purification of a cold-sensitive microsomal enzyme. *J Biol Chem* 1973;248: pp. 4731–8.

[85] Brown MS, Dana SE, Goldstein JL. Receptor-dependent hydrolysis of cholesteryl esters contained in plasma low density lipoprotein. *Proc Natl Acad Sci U S A* 1975;72: pp. 2925–9.

[86] Brown MS, Dana SE, Goldstein JL. Cholesterol ester formation in cultured human fibroblasts. Stimulation by oxygenated sterols. *J Biol Chem* 1975;250: pp. 4025–7.

[87] Brown MS, Dana SE, Goldstein JL. Regulation of 3-hydroxy-3-methylglutaryl coenzyme

A reductase activity in cultured human fibroblasts. Comparison of cells from a normal subject and from a patient with homozygous familial hypercholesterolemia. *J Biol Chem* 1974;249: pp. 789–96.

[88] Brown MS, Dana SE, Goldstein JL. Regulation of 3-hydroxy-3-methylglutaryl coenzyme A reductase activity in human fibroblasts by lipoproteins. *Proc Natl Acad Sci U S A* 1973;70: pp. 2162–6.

[89] Dietschy JM, Spady DK. Measurement of rates of cholesterol synthesis using tritiated water. *J Lipid Res* 1984;25: pp. 1469–76.

[90] Dietschy JM, Spady DK. Regulation of low density lipoprotein uptake and degradation in different animals species. *Agents Actions Suppl* 1984;16: pp. 177–90.

[91] Dietschy JM, Spady DK, Stange EF. Quantitative importance of different organs for cholesterol synthesis and low-density-lipoprotein degradation. *Biochem Soc Trans* 1983;11: pp. 639–41.

[92] Dietschy JM, Turley SD, Spady DK. Role of liver in the maintenance of cholesterol and low density lipoprotein homeostasis in different animal species, including humans. *J Lipid Res* 1993;34: pp. 1637–59.

[93] Turley SD, Spady DK, Dietschy JM. Identification of a metabolic difference accounting for the hyper- and hyporesponder phenotypes of cynomolgus monkey. *J Lipid Res* 1997;38: pp. 1598–611.

[94] Turley SD, Spady DK, Dietschy JM. Regulation of fecal bile acid excretion in male golden Syrian hamsters fed a cereal-based diet with and without added cholesterol. *Hepatology* 1997;25: pp. 797–803.

[95] Turley SD, Spady DK, Dietschy JM. Role of liver in the synthesis of cholesterol and the clearance of low density lipoproteins in the cynomolgus monkey. *J Lipid Res* 1995;36: pp. 67–79.

[96] Spady DK, Dietschy JM. Sterol synthesis in vivo in 18 tissues of the squirrel monkey, guinea pig, rabbit, hamster, and rat. *J Lipid Res* 1983;24: pp. 303–15.

[97] Yu L, Gupta S, Xu F, Liverman AD, Moschetta A, Mangelsdorf DJ, Repa JJ, et al. Expression of ABCG5 and ABCG8 is required for regulation of biliary cholesterol secretion. *J Biol Chem* 2005;280: pp. 8742–7.

[98] Yu L, Hammer RE, Li-Hawkins J, Von Bergmann K, Lutjohann D, Cohen JC, Hobbs HH. Disruption of Abcg5 and Abcg8 in mice reveals their crucial role in biliary cholesterol secretion. *Proc Natl Acad Sci U S A* 2002;99: pp. 16237–42.

[99] Klett EL, Lu K, Kosters A, Vink E, Lee MH, Altenburg M, Shefer S, et al. A mouse model of sitosterolemia: absence of Abcg8/sterolin-2 results in failure to secrete biliary cholesterol. *BMC Med* 2004;2: p. 5.

[100] Kosters A, Frijters RJ, Schaap FG, Vink E, Plosch T, Ottenhoff R, Jirsa M, et al. Relation

between hepatic expression of ATP-binding cassette transporters G5 and G8 and biliary cholesterol secretion in mice. *J Hepatol* 2003;38: pp. 710–6.

[101] Wang HH, Lammert F, Schmitz A, Wang DQ. Transgenic overexpression of Abcb11 enhances biliary bile salt outputs, but does not affect cholesterol cholelithogenesis in mice. *Eur J Clin Invest* 2010;40: pp. 541–51.

[102] Admirand WH, Small DM. The physicochemical basis of cholesterol gallstone formation in man. *J Clin Invest* 1968;47: pp. 1043–52.

[103] Carey MC, Small DM. The physical chemistry of cholesterol solubility in bile. Relationship to gallstone formation and dissolution in man. *J Clin Invest* 1978;61: pp. 998–1026.

[104] Wang DQ, Carey MC. Characterization of crystallization pathways during cholesterol precipitation from human gallbladder biles: identical pathways to corresponding model biles with three predominating sequences. *J Lipid Res* 1996;37: pp. 2539–49.

[105] Wang DQ, Cohen DE, Lammert F, Carey MC. No pathophysiologic relationship of soluble biliary proteins to cholesterol crystallization in human bile. *J Lipid Res* 1999;40: pp. 415–25.

[106] Brown MS, Goldstein JL. How LDL receptors influence cholesterol and atherosclerosis. *Sci Am* 1984;251: pp. 58–66.

[107] Brown MS, Goldstein JL. Lipoprotein metabolism in the macrophage: implications for cholesterol deposition in atherosclerosis. *Annu Rev Biochem* 1983;52: pp. 223–61.

[108] Small DM. Cellular mechanisms for lipid deposition in atherosclerosis (first of two parts). *N Engl J Med* 1977;297: pp. 873–7.

[109] Small DM. George Lyman Duff memorial lecture. Progression and regression of atherosclerotic lesions. Insights from lipid physical biochemistry. *Arteriosclerosis* 1988;8: pp. 103–29.

[110] Small DM, Shipley GG. Physical–chemical basis of lipid deposition in atherosclerosis. *Science* 1974;185: pp. 222–9.

[111] Goldstein JL, Brown MS. Regulation of low-density lipoprotein receptors: implications for pathogenesis and therapy of hypercholesterolemia and atherosclerosis. *Circulation* 1987;76: pp. 504–7.

[112] Portincasa P, Moschetta A, Di Ciaula A, Pontrelli D, Sasso RC, Wang HH, Wang DQ: Pathophysiology and Cholesterol Gallstone Disease. In: Borzellino G, Cordiano C, eds. Biliary Lithiasis: Basic Science, Current Diagnosis and Management. 1 ed. Milano: Springer Italia S.r.l., 2008; pp. 19–49.

[113] Portincasa P, Moschetta A, Palasciano G. Cholesterol gallstone disease. *Lancet* 2006;368: pp. 230–9.

[114] Portincasa P, Moschetta A, Palasciano G. From lipid secretion to cholesterol crystallization in bile. Relevance in cholesterol gallstone disease. *Ann Hepatol* 2002;1: pp. 121–8.

[115] Portincasa P, Moschetta A, van Erpecum KJ, Calamita G, Margari A, vanBerge-Henegouwen GP, Palasciano G. Pathways of cholesterol crystallization in model bile and native bile. *Dig Liver Dis* 2003;35: pp. 118–26.

[116] Afdhal NH, Smith BF. Cholesterol crystal nucleation: a decade-long search for the missing link in gallstone pathogenesis. *Hepatology* 1990;11: pp. 699–702.

[117] Holan KR, Holzbach RT, Hermann RE, Cooperman AM, Claffey WJ. Nucleation time: a key factor in the pathogenesis of cholesterol gallstone disease. *Gastroenterology* 1979;77: pp. 611–7.

[118] Holzbach RT. Cholesterol nucleation in bile. *Ital J Gastroenterol* 1995;27: pp. 101–5.

[119] Holzbach RT. Nucleation of cholesterol crystals in native bile. *Hepatology* 1990;12: 155S–9S; discussion 159S–61S.

[120] Holzbach RT. Recent progress in understanding cholesterol crystal nucleation as a precursor to human gallstone formation. *Hepatology* 1986;6: pp. 1403–6.

[121] Holzbach RT. Factors influencing cholesterol nucleation in bile. *Hepatology* 1984;4: pp. 173S–6S.

[122] Holzbach RT, Busch N. Nucleation and growth of cholesterol crystals. Kinetic determinants in supersaturated native bile. *Gastroenterol Clin North Am* 1991;20: pp. 67–84.

[123] Jeske DJ, Dietschy JM. Regulation of rates of cholesterol synthesis in vivo in the liver and carcass of the rat measured using [3H]water. *J Lipid Res* 1980;21: pp. 364–76.

[124] Andersen JM, Dietschy JM. Absolute rates of cholesterol synthesis in extrahepatic tissues measured with 3H-labeled water and 14C-labeled substrates. *J Lipid Res* 1979;20: pp. 740–52.

[125] Dietschy JM, Kita T, Suckling KE, Goldstein JL, Brown MS. Cholesterol synthesis in vivo and in vitro in the WHHL rabbit, an animal with defective low density lipoprotein receptors. *J Lipid Res* 1983;24: pp. 469–80.

[126] Dietschy JM, McGarry JD. Limitations of acetate as a substrate for measuring cholesterol synthesis in liver. *J Biol Chem* 1974;249: pp. 52–8.

[127] Spady DK, Dietschy JM. Dietary saturated triacylglycerols suppress hepatic low density lipoprotein receptor activity in the hamster. *Proc Natl Acad Sci U S A* 1985;82: pp. 4526–30.

[128] Spady DK, Huettinger M, Bilheimer DW, Dietschy JM. Role of receptor-independent low density lipoprotein transport in the maintenance of tissue cholesterol balance in the normal and WHHL rabbit. *J Lipid Res* 1987;28: pp. 32–41.

[129] Grundy SM, Ahrens EH, Jr. Measurements of cholesterol turnover, synthesis, and absorption in man, carried out by isotope kinetic and sterol balance methods. *J Lipid Res* 1969;10: pp. 91–107.

[130] Grundy SM, Ahrens EH, Jr., Davignon J. The interaction of cholesterol absorption and cholesterol synthesis in man. *J Lipid Res* 1969;10: pp. 304–15.

[131] Grundy SM, Ahrens EH, Jr., Salen G. Dietary beta-sitosterol as an internal standard to correct for cholesterol losses in sterol balance studies. *J Lipid Res* 1968;9: pp. 374–87.

[132] Quintao E, Grundy SM, Ahrens EH, Jr. Effects of dietary cholesterol on the regulation of total body cholesterol in man. *J Lipid Res* 1971;12: pp. 233–47.

[133] Quintao E, Grundy SM, Ahrens EH, Jr. An evaluation of four methods for measuring cholesterol absorption by the intestine in man. *J Lipid Res* 1971;12: pp. 221–32.

[134] Andersen JM, Dietschy JM. Relative importance of high and low density lipoproteins in the regulation of cholesterol synthesis in the adrenal gland, ovary, and testis of the rat. *J Biol Chem* 1978;253: pp. 9024–32.

[135] Goldstein JL, Brown MS. Regulation of the mevalonate pathway. *Nature* 1990;343: pp. 425–30.

[136] Horton JD, Goldstein JL, Brown MS. SREBPs: activators of the complete program of cholesterol and fatty acid synthesis in the liver. *J Clin Invest* 2002;109: pp. 1125–31.

[137] Horton JD, Shah NA, Warrington JA, Anderson NN, Park SW, Brown MS, Goldstein JL. Combined analysis of oligonucleotide microarray data from transgenic and knockout mice identifies direct SREBP target genes. *Proc Natl Acad Sci U S A* 2003;100: pp. 12027–32.

[138] Landau JM, Sekowski A, Hamm MW. Dietary cholesterol and the activity of stearoyl CoA desaturase in rats: evidence for an indirect regulatory effect. *Biochim Biophys Acta* 1997;1345: pp. 349–57.

[139] Osborne TF. Sterol regulatory element-binding proteins (SREBPs): key regulators of nutritional homeostasis and insulin action. *J Biol Chem* 2000;275: pp. 32379–82.

[140] Sakakura Y, Shimano H, Sone H, Takahashi A, Inoue N, Toyoshima H, Suzuki S, et al. Sterol regulatory element-binding proteins induce an entire pathway of cholesterol synthesis. *Biochem Biophys Res Commun* 2001;286: pp. 176–83.

[141] McKenney JM, Ganz P, Wiggins BS, Saseen JS: Statins. In: Ballantyne CM, ed. Lipidology in the Treatment and Prevention of Cardiovascular Disease (Clinical Lipidology: A Companion to Braunwald's Heart Disease). 1 ed. Philadelphia: Elsevier Saunders, 2008; p. 253.

[142] Brown MS, Goldstein JL. The SREBP pathway: regulation of cholesterol metabolism by proteolysis of a membrane-bound transcription factor. *Cell* 1997;89: pp. 331–40.

[143] Goldstein JL, Rawson RB, Brown MS. Mutant mammalian cells as tools to delineate the sterol regulatory element-binding protein pathway for feedback regulation of lipid synthesis. *Arch Biochem Biophys* 2002;397: pp. 139–48.

[144] Brown MS, Goldstein JL. Sterol regulatory element binding proteins (SREBPs): controllers of lipid synthesis and cellular uptake. *Nutr Rev* 1998;56:S1–3; discussion S54–75.

[145] Hua X, Nohturfft A, Goldstein JL, Brown MS. Sterol resistance in CHO cells traced to point mutation in SREBP cleavage-activating protein. *Cell* 1996;87: pp. 415–26.

[146] Edwards PA, Tabor D, Kast HR, Venkateswaran A. Regulation of gene expression by SREBP and SCAP. *Biochim Biophys Acta* 2000;1529: pp. 103–13.

[147] Sato R, Inoue J, Kawabe Y, Kodama T, Takano T, Maeda M. Sterol-dependent transcriptional regulation of sterol regulatory element-binding protein-2. *J Biol Chem* 1996;271: pp. 26461–4.

[148] Hofmann AF, Sjovall J, Kurz G, Radominska A, Schteingart CD, Tint GS, Vlahcevic ZR, et al. A proposed nomenclature for bile acids. *J Lipid Res* 1992;33: pp. 599–604.

[149] Hofmann AF, Hagey LR, Krasowski MD. Bile salts of vertebrates: structural variation and possible evolutionary significance. *J Lipid Res* 2010;51: pp. 226–46.

[150] Hofmann AF: Bile Acids. In: Arias IM, Jakoby WB, Popper H, Schachter D, Shafritz, DA, eds. The Liver: Biology and Pathobiology. 2 ed. New York: Raven Press, 1988; pp. 553–72.

[151] Roda A, Hofmann AF, Mysels KJ. The influence of bile salt structure on self-association in aqueous solutions. *J Biol Chem* 1983;258: pp. 6362–70.

[152] Hofmann AF, Mysels KJ. Bile acid solubility and precipitation in vitro and in vivo: the role of conjugation, pH, and Ca2+ ions. *J Lipid Res* 1992;33: pp. 617–26.

[153] Hofmann AF. Ionized calcium in gallbladder bile: new insights into a clear and ever-present danger. *J Lab Clin Med* 1992;120: pp. 818–20.

[154] Carey MC. Bile salt structure and phase equilibria in aqueous bile salt and bile salt-lecithin systems. *Hepatology* 1984;4: pp. 138S–42S.

[155] Carey MC, Small DM. Micellar properties of dihydroxy and trihydroxy bile salts: effects of counterion and temperature. *J Colloid Interface Sci* 1969;31: pp. 382–96.

[156] Carey MC. Bile acids and bile salts: ionization and solubility properties. *Hepatology* 1984;4: pp. 66S–71S.

[157] Carey MC, Montet JC, Phillips MC, Armstrong MJ, Mazer NA. Thermodynamic and molecular basis for dissimilar cholesterol-solubilizing capacities by micellar solutions of bile salts: cases of sodium chenodeoxycholate and sodium ursodeoxycholate and their glycine and taurine conjugates. *Biochemistry* 1981;20: pp. 3637–48.

[158] Carey MC, Small DM. Micelle formation by bile salts. Physical–chemical and thermodynamic considerations. *Arch Intern Med* 1972;130: pp. 506–27.

[159] Angelico M, De Sanctis SC, Gandin C, Alvaro D. Spontaneous formation of pigmentary precipitates in bile salt-depleted rat bile and its prevention by micelle-forming bile salts. *Gastroenterology* 1990;98: pp. 444–53.

[160] Carey MC, Small DM. Micellar properties of sodium fusidate, a steroid antibiotic structurally resembling the bile salts. *J Lipid Res* 1971;12: pp. 604–13.

[161] Carey MC, Wu SF, Watkins JB. Solution properties of sulfated monohydroxy bile salts. Relative insolubility of the disodium salt of glycolithocholate sulfate. *Biochim Biophys Acta* 1979;575: pp. 16–26.

[162] Cohen DE, Carey MC. Physical chemistry of biliary lipids during bile formation. *Hepatology* 1990;12:143S–7S; discussion 147S–8S.

[163] Roda A, Grigolo B, Aldini R, Simoni P, Pellicciari R, Natalini B, Balducci R. Bile acids with a cyclopropyl-containing side chain. IV. Physicochemical and biological properties of the four diastereoisomers of 3 alpha,7 beta-dihydroxy-22,23-methylene-5 beta-cholan-24-oic acid. *J Lipid Res* 1987;28: pp. 1384–97.

[164] Roda A, Grigolo B, Minutello A, Pellicciari R, Natalini B. Physicochemical and biological properties of natural and synthetic C-22 and C-23 hydroxylated bile acids. *J Lipid Res* 1990;31: pp. 289–98.

[165] Hofmann AF, Roda A. Physicochemical properties of bile acids and their relationship to biological properties: an overview of the problem. *J Lipid Res* 1984;25: pp. 1477–89.

[166] Apstein MD, Carey MC. Pathogenesis of cholesterol gallstones: a parsimonious hypothesis. *Eur J Clin Invest* 1996;26: pp. 343–52.

[167] Armstrong MJ, Carey MC. Thermodynamic and molecular determinants of sterol solubilities in bile salt micelles. *J Lipid Res* 1987;28: pp. 1144–55.

[168] Cohen DE, Angelico M, Carey MC. Structural alterations in lecithin-cholesterol vesicles following interactions with monomeric and micellar bile salts: physical–chemical basis for subselection of biliary lecithin species and aggregative states of biliary lipids during bile formation. *J Lipid Res* 1990;31: pp. 55–70.

[169] Cohen DE, Angelico M, Carey MC. Quasielastic light scattering evidence for vesicular secretion of biliary lipids. *Am J Physiol* 1989;257: pp. G1–8.

[170] Cohen DE, Carey MC. Acyl chain unsaturation modulates distribution of lecithin molecular species between mixed micelles and vesicles in model bile. Implications for particle structure and metastable cholesterol solubilities. *J Lipid Res* 1991;32: pp. 1291–302.

[171] Carey MC, Mazer NA. Biliary lipid secretion in health and in cholesterol gallstone disease. *Hepatology* 1984;4: pp. 31S–7S.

[172] Cai X, Grant DJ, Wiedmann TS. Analysis of the solubilization of steroids by bile salt micelles. *J Pharm Sci* 1997;86: pp. 372–7.

[173] Hofmann AF. The continuing importance of bile acids in liver and intestinal disease. *Arch Intern Med* 1999;159: pp. 2647–58.

[174] Batta AK, Salen G, Arora R, Shefer S, Batta M, Person A. Side chain conjugation prevents bacterial 7-dehydroxylation of bile acids. *J Biol Chem* 1990;265: pp. 10925–8.

[175] Clayton LM, Gurantz D, Hofmann AF, Hagey LR, Schteingart CD. Role of bile acid conjugation in hepatic transport of dihydroxy bile acids. *J Pharmacol Exp Ther* 1989;248: pp. 1130–7.

[176] Fischer S, Neubrand M, Paumgartner G. Biotransformation of orally administered urso-

deoxycholic acid in man as observed in gallbladder bile, serum and urine. *Eur J Clin* Invest 1993;23: pp. 28–36.

[177] Aldini R, Roda A, Montagnani M, Roda E. Bile acid structure and intestinal absorption in the animal model. *Ital J Gastroenterol* 1995;27: pp. 141–4.

[178] Alnouti Y. Bile Acid sulfation: a pathway of bile acid elimination and detoxification. *Toxicol Sci* 2009;108: pp. 225–46.

[179] Hofmann AF. Bile acid secretion, bile flow and biliary lipid secretion in humans. *Hepatology* 1990;12:17S–22S; discussion 22S–5S.

[180] Hofmann AF, Strandvik B. Defective bile acid amidation: predicted features of a new inborn error of metabolism. *Lancet* 1988;2: pp. 311–3.

[181] Yoon YB, Hagey LR, Hofmann AF, Gurantz D, Michelotti EL, Steinbach JH. Effect of side-chain shortening on the physiologic properties of bile acids: hepatic transport and effect on biliary secretion of 23-nor-ursodeoxycholate in rodents. *Gastroenterology* 1986;90: pp. 837–52.

[182] Hofmann AF. The enterohepatic circulation of conjugated bile acids in healthy man: quantitative description and functions. *Expos Annu Biochim Med* 1977;33: pp. 69–86.

[183] Hofmann AF. The enterohepatic circulation of bile acids in man. *Clin Gastroenterol* 1977;6: pp. 3–24.

[184] Vlahcevic ZR, Gurley EC, Heuman DM, Hylemon PB. Bile salts in submicellar concentrations promote bidirectional cholesterol transfer (exchange) as a function of their hydrophobicity. *J Lipid Res* 1990;31: pp. 1063–71.

[185] Heuman DM. Distribution of mixtures of bile salt taurine conjugates between lecithin-cholesterol vesicles and aqueous media: an empirical model. *J Lipid Res* 1997;38: pp. 1217–28.

[186] Heuman DM. Quantitative estimation of the hydrophilic-hydrophobic balance of mixed bile salt solutions. *J Lipid Res* 1989;30: pp. 719–30.

[187] Wang DQ, Tazuma S, Cohen DE, Carey MC. Feeding natural hydrophilic bile acids inhibits intestinal cholesterol absorption: studies in the gallstone-susceptible mouse. *Am J Physiol Gastrointest Liver Physiol* 2003;285: pp. G494–502.

[188] Salvioli G, Igimi H, Carey MC. Cholesterol gallstone dissolution in bile. Dissolution kinetics of crystalline cholesterol monohydrate by conjugated chenodeoxycholate-lecithin and conjugated ursodeoxycholate–lecithin mixtures: dissimilar phase equilibria and dissolution mechanisms. *J Lipid Res* 1983;24: pp. 701–20.

[189] Wang DQ, Carey MC. Complete mapping of crystallization pathways during cholesterol precipitation from model bile: influence of physical–chemical variables of pathophysiologic relevance and identification of a stable liquid crystalline state in cold, dilute and hydrophilic bile salt-containing systems. *J Lipid Res* 1996;37: pp. 606–30.

[190] Cabral DJ, Hamilton JA, Small DM. The ionization behavior of bile acids in different aqueous environments. *J Lipid Res* 1986;27: pp. 334–43.

[191] Armstrong MJ, Carey MC. The hydrophobic-hydrophilic balance of bile salts. Inverse correlation between reverse-phase high performance liquid chromatographic mobilities and micellar cholesterol-solubilizing capacities. *J Lipid Res* 1982;23: pp. 70–80.

[192] Attili AF, Angelico M, Cantafora A, Alvaro D, Capocaccia L. Bile acid-induced liver toxicity: relation to the hydrophobic-hydrophilic balance of bile acids. *Med Hypotheses* 1986;19: pp. 57–69.

[193] Bilhartz LE, Dietschy JM. Bile salt hydrophobicity influences cholesterol recruitment from rat liver in vivo when cholesterol synthesis and lipoprotein uptake are constant. *Gastroenterology* 1988;95: pp. 771–9.

[194] Cohen DE, Leighton LS, Carey MC. Bile salt hydrophobicity controls vesicle secretion rates and transformations in native bile. *Am J Physiol* 1992;263: pp. G386–95.

[195] Fickert P, Zollner G, Fuchsbichler A, Stumptner C, Pojer C, Zenz R, Lammert F, et al. Effects of ursodeoxycholic and cholic acid feeding on hepatocellular transporter expression in mouse liver. *Gastroenterology* 2001;121: pp. 170–83.

[196] Trauner M, Fickert P, Halilbasic E, Moustafa T. Lessons from the toxic bile concept for the pathogenesis and treatment of cholestatic liver diseases. *Wien Med Wochenschr* 2008;158: pp. 542–8.

[197] Van de Meeberg PC, van Erpecum KJ, van Berge-Henegouwen GP. Therapy with ursodeoxycholic acid in cholestatic liver disease. *Scand J Gastroenterol Suppl* 1993;200: pp. 15–20.

[198] van de Meeberg PC, Wolfhagen FH, van Erpecum KJ, van Berge Henegouwen GP. Cholestatic liver diseases: new strategies for prevention and treatment of hepatobiliary and cholestatic diseases. *Neth J Med* 1995;47: pp. 30–5.

[199] Rost D, Herrmann T, Sauer P, Schmidts HL, Stieger B, Meier PJ, Stremmel W, et al. Regulation of rat organic anion transporters in bile salt-induced cholestatic hepatitis: effect of ursodeoxycholate. *Hepatology* 2003;38: pp. 187–95.

[200] Salen G, Batta AK. Bile acid abnormalities in cholestatic liver diseases. *Gastroenterol Clin North Am* 1999;28: pp. 173–93.

[201] Serrano MA, Macias RI, Vallejo M, Briz O, Bravo A, Pascual MJ, St-Pierre MV, et al. Effect of ursodeoxycholic acid on the impairment induced by maternal cholestasis in the rat placenta-maternal liver tandem excretory pathway. *J Pharmacol Exp Ther* 2003;305: pp. 515–24.

[202] Bellentani S. Immunomodulating and anti-apoptotic action of ursodeoxycholic acid: where are we and where should we go? *Eur J Gastroenterol Hepatol* 2005;17: pp. 137–40.

[203] Cohen BI, Hofmann AF, Mosbach EH, Stenger RJ, Rothschild MA, Hagey LR, Yoon YB.

Differing effects of nor-ursodeoxycholic or ursodeoxycholic acid on hepatic histology and bile acid metabolism in the rabbit. *Gastroenterology* 1986;91: pp. 189–97.

[204] Crosignani A, Battezzati PM, Setchell KD, Camisasca M, Bertolini E, Roda A, Zuin M, et al. Effects of ursodeoxycholic acid on serum liver enzymes and bile acid metabolism in chronic active hepatitis: a dose-response study. *Hepatology* 1991;13: pp. 339–44.

[205] Festi D, Montagnani M, Azzaroli F, Lodato F, Mazzella G, Roda A, Di Biase AR, et al. Clinical efficacy and effectiveness of ursodeoxycholic acid in cholestatic liver diseases. *Curr Clin Pharmacol* 2007;2: pp. 155–77.

[206] Fracchia M, Setchell KD, Crosignani A, Podda M, O'Connell N, Ferraris R, Hofmann AF, et al. Bile acid conjugation in early stage cholestatic liver disease before and during treatment with ursodeoxycholic acid. *Clin Chim Acta* 1996;248: pp. 175–85.

[207] Trauner M, Graziadei IW. Review article: mechanisms of action and therapeutic applications of ursodeoxycholic acid in chronic liver diseases. *Aliment Pharmacol Ther* 1999;13: pp. 979–96.

[208] Vacca M, Krawczyk M, Petruzzelli M, Sasso RC, van Erpecum KJ, Palasciano G, van Berge-Henegouwen GP, et al. Current treatments of primary sclerosing cholangitis. *Curr Med Chem* 2007;14: pp. 2081–94.

[209] Paumgartner G. Biliary physiology and disease: reflections of a physician-scientist. *Hepatology* 2010;51: pp. 1095–106.

[210] Podda M, Ghezzi C, Battezzati PM, Bertolini E, Crosignani A, Petroni ML, Zuin M. Effect of different doses of ursodeoxycholic acid in chronic liver disease. *Dig Dis Sci* 1989;34: pp. 59S–65S.

[211] Aggarwal SK, Batta AK, Salen G, Shefer S. Synthesis of 3 alpha,6 beta,7 alpha,12 beta- and 3 alpha,6 beta,7 beta,12 beta-tetrahydroxy-5 beta-cholanoic acids. *Steroids* 1992;57: pp. 107–11.

[212] Vlahcevic ZR, Pandak WM, Heuman DM, Hylemon PB. Function and regulation of hydroxylases involved in the bile acid biosynthesis pathways. *Semin Liver Dis* 1992;12: pp. 403–19.

[213] Vlahcevic ZR. Regulation of cholesterol 7 alpha-hydroxylase by different effectors. *Ital J Gastroenterol* 1996;28: pp. 337–9.

[214] Vlahcevic ZR, Bell CC, Jr., Buhac I, Farrar JT, Swell L. Diminished bile acid pool size in patients with gallstones. *Gastroenterology* 1970;59: pp. 165–73.

[215] Vlahcevic ZR, Cronholm T, Curstedt T, Sjovall J. Biosynthesis of 5 alpha- and 5 beta-cholanoic acid derivatives during metabolism of [1,1-2H]- and [2,2,2-2H]ethanol in the rat. *Biochim Biophys Acta* 1980;618: pp. 369–77.

[216] Vlahcevic ZR, Eggertsen G, Bjorkhem I, Hylemon PB, Redford K, Pandak WM. Regulation of sterol 12alpha-hydroxylase and cholic acid biosynthesis in the rat. *Gastroenterology* 2000;118: pp. 599–607.

[217] Vlahcevic ZR, Goldman M, Schwartz CC, Gustafsson J, Swell L. Bile acid metabolism in cirrhosis. VII. Evidence for defective feedback control of bile acid synthesis. *Hepatology* 1981;1: pp. 146–50.

[218] Vlahcevic ZR, Heuman DM, Hylemon PB. Regulation of bile acid synthesis. *Hepatology* 1991;13: pp. 590–600.

[219] Vlahcevic ZR, Jairath SK, Heuman DM, Stravitz RT, Hylemon PB, Avadhani NG, Pandak WM. Transcriptional regulation of hepatic sterol 27-hydroxylase by bile acids. *Am J Physiol* 1996;270: pp. G646–52.

[220] Vlahcevic ZR, Juttijudata P, Bell CC, Jr., Swell L. Bile acid metabolism in patients with cirrhosis. II. Cholic and chenodeoxycholic acid metabolism. *Gastroenterology* 1972;62: pp. 1174–81.

[221] Vlahcevic ZR, Miller JR, Farrar JT, Swell L. Kinetics and pool size of primary bile acids in man. *Gastroenterology* 1971;61: pp. 85–90.

[222] Vlahcevic ZR, Pandak WM, Hylemon PB, Heuman DM. Role of newly synthesized cholesterol or its metabolites on the regulation of bile acid biosynthesis after short-term biliary diversion in the rat. *Hepatology* 1993;18: pp. 660–8.

[223] Vlahcevic ZR, Pandak WM, Stravitz RT. Regulation of bile acid biosynthesis. *Gastroenterol Clin North Am* 1999;28: pp. 1–25, v.

[224] Vlahcevic ZR, Prugh MF, Gregory DH, Swell L. Disturbances of bile acid metabolism in parenchymal liver cell disease. *Clin Gastroenterol* 1977;6: pp. 25–43.

[225] Vlahcevic ZR, Schwartz CC, Gustafsson J, Halloran LG, Danielsson H, Swell L. Biosynthesis of bile acids in man. Multiple pathways to cholic acid and chenodeoxycholic acid. *J Biol Chem* 1980;255: pp. 2925–33.

[226] Heuman DM, Hernandez CR, Hylemon PB, Kubaska WM, Hartman C, Vlahcevic ZR. Regulation of bile acid synthesis. I. Effects of conjugated ursodeoxycholate and cholate on bile acid synthesis in chronic bile fistula rat. *Hepatology* 1988;8: pp. 358–65.

[227] Heuman DM, Hylemon PB, Vlahcevic ZR. Regulation of bile acid synthesis. III. Correlation between biliary bile salt hydrophobicity index and the activities of enzymes regulating cholesterol and bile acid synthesis in the rat. *J Lipid Res* 1989;30: pp. 1161–71.

[228] Heuman DM, Vlahcevic ZR, Bailey ML, Hylemon PB. Regulation of bile acid synthesis. II. Effect of bile acid feeding on enzymes regulating hepatic cholesterol and bile acid synthesis in the rat. *Hepatology* 1988;8: pp. 892–7.

[229] Carey MC, Cahalane MJ: Enterohepatic Circulation. In: Arias IM, Jakoby WB, Popper H,

Schachter D, Shafritz, DA, eds. The Liver: Biology and Pathobiology. 2 ed. New York: Raven Press, 1988; pp. 573–616.

[230] Alberts DS, Einspahr JG, Earnest DL, Krutzsch MF, Lin P, Hess LM, Heddens DK, et al. Fecal bile acid concentrations in a subpopulation of the wheat bran fiber colon polyp trial. *Cancer Epidemiol Biomarkers Prev* 2003;12: pp. 197–200.

[231] Faloon WW, Rubulis A, Knipp J, Sherman CD, Flood MS. Fecal fat, bile acid, and sterol excretion abd biliary lipid changes in jejunoileostomy patients. *Am J Clin Nutr* 1977;30: pp. 21–31.

[232] Hamilton JP, Xie G, Raufman JP, Hogan S, Griffin TL, Packard CA, Chatfield DA, et al. Human cecal bile acids: concentration and spectrum. *Am J Physiol Gastrointest Liver Physiol* 2007;293: pp. G256–63.

[233] Lee C, Martin KO, Javitt NB. Bile acid synthesis: 7 alpha-hydroxylation of intermediates in the sterol 27-hydroxylase metabolic pathway. *J Lipid Res* 1996;37: pp. 1356–62.

[234] Honda A, Salen G, Shefer S, Matsuzaki Y, Xu G, Batta AK, Tint GS, et al. Regulation of 25- and 27-hydroxylation side chain cleavage pathways for cholic acid biosynthesis in humans, rabbits, and mice. Assay of enzyme activities by high-resolution gas chromatography;-mass spectrometry. *J Lipid Res* 2000;41: pp. 442–51.

[235] Dubrac S, Lear SR, Ananthanarayanan M, Balasubramaniyan N, Bollineni J, Shefer S, Hyogo H, et al. Role of CYP27A in cholesterol and bile acid metabolism. *J Lipid Res* 2005;46: pp. 76–85.

[236] Okuda K, Masumoto O, Ohyama Y. Purification and characterization of 5 beta-cholestane-3 alpha,7 alpha,12 alpha-triol 27-hydroxylase from female rat liver mitochondria. *J Biol Chem* 1988;263: pp. 18138–42.

[237] Pandak WM, Hylemon PB, Ren S, Marques D, Gil G, Redford K, Mallonee D, et al. Regulation of oxysterol 7alpha-hydroxylase (CYP7B1) in primary cultures of rat hepatocytes. *Hepatology* 2002;35: pp. 1400–8.

[238] Petrack B, Latario BJ. Synthesis of 27-hydroxycholesterol in rat liver mitochondria: HPLC assay and marked activation by exogenous cholesterol. *J Lipid Res* 1993;34: pp. 643–9.

[239] Ueki I, Kimura A, Nishiyori A, Chen HL, Takei H, Nittono H, Kurosawa T. Neonatal cholestatic liver disease in an Asian patient with a homozygous mutation in the oxysterol 7alpha-hydroxylase gene. *J Pediatr Gastroenterol Nutr* 2008;46: pp. 465–9.

[240] Setchell KD, Schwarz M, O'Connell NC, Lund EG, Davis DL, Lathe R, Thompson HR, et al. Identification of a new inborn error in bile acid synthesis: mutation of the oxysterol 7alpha-hydroxylase gene causes severe neonatal liver disease. *J Clin Invest* 1998;102: pp. 1690–703.

[241] Arnoldi A, Crimella C, Tenderini E, Martinuzzi A, D'Angelo M, Musumeci O, Toscano

A, et al. Clinical phenotype variability in patients with hereditary spastic paraplegia type 5 associated with CYP7B1 mutations. *Clin Genet* 2012;81: pp. 150–7.

[242] Xie C, Lund EG, Turley SD, Russell DW, Dietschy JM. Quantitation of two pathways for cholesterol excretion from the brain in normal mice and mice with neurodegeneration. *J Lipid Res* 2003;44: pp. 1780–9.

[243] Lund EG, Xie C, Kotti T, Turley SD, Dietschy JM, Russell DW. Knockout of the cholesterol 24-hydroxylase gene in mice reveals a brain-specific mechanism of cholesterol turnover. *J Biol Chem* 2003;278: pp. 22980–8.

[244] Javitt NB. Cholesterol, hydroxycholesterols, and bile acids. *Biochem Biophys Res Commun* 2002;292: pp. 1147–53.

[245] Javitt NB. 25R,26-Hydroxycholesterol revisited: synthesis, metabolism, and biologic roles. *J Lipid Res* 2002;43: pp. 665–70.

[246] Ren S, Marques D, Redford K, Hylemon PB, Gil G, Vlahcevic ZR, Pandak WM. Regulation of oxysterol 7alpha-hydroxylase (CYP7B1) in the rat. *Metabolism* 2003;52: pp. 636–42.

[247] Alrefai WA, Gill RK. Bile acid transporters: structure, function, regulation and pathophysiological implications. *Pharm Res* 2007;24: pp. 1803–23.

[248] Duane WC, Hartich LA, Bartman AE, Ho SB. Diminished gene expression of ileal apical sodium bile acid transporter explains impaired absorption of bile acid in patients with hypertriglyceridemia. *J Lipid Res* 2000;41: pp. 1384–9.

[249] Xu G, Shneider BL, Shefer S, Nguyen LB, Batta AK, Tint GS, Arrese M, et al. Ileal bile acid transport regulates bile acid pool, synthesis, and plasma cholesterol levels differently in cholesterol-fed rats and rabbits. *J Lipid Res* 2000;41: pp. 298–304.

[250] Ho RH, Leake BF, Urquhart BL, Gregor JC, Dawson PA, Kim RB. Functional characterization of genetic variants in the apical sodium-dependent bile acid transporter (ASBT; SLC10A2). *J Gastroenterol Hepatol* 2011.

[251] Angelin B, Nilsell K, Einarsson K. Ursodeoxycholic acid treatment in humans: effects on plasma and biliary lipid metabolism with special reference to very low density lipoprotein triglyceride and bile acid kinetics. *Eur J Clin Invest* 1986;16: pp. 169–77.

[252] Fedorowski T, Salen G, Calallilo A, Tint GS, Mosbach EH, Hall JC. Metabolism of ursodeoxycholic acid in man. *Gastroenterology* 1977;73: pp. 1131–7.

[253] Bachrach WH, Hofmann AF. Ursodeoxycholic acid in the treatment of cholesterol cholelithiasis. part I. *Dig Dis Sci* 1982;27: pp. 737–61.

[254] Bachrach WH, Hofmann AF. Ursodeoxycholic acid in the treatment of cholesterol cholelithiasis. Part II. *Dig Dis Sci* 1982;27: pp. 833–56.

[255] Cowen AE, Korman MG, Hofmann AF, Cass OW. Metabolism of lethocholate in healthy

man. I. Biotransformation and biliary excretion of intravenously administered lithocholate, lithocholylglycine, and their sulfates. *Gastroenterology* 1975;69: pp. 59–66.

[256] Cowen AE, Korman MG, Hofmann AF, Cass OW, Coffin SB. Metabolism of lithocholate in healthy man. II. Enterohepatic circulation. *Gastroenterology* 1975;69: pp. 67–76.

[257] Cowen AE, Korman MG, Hofmann AF, Thomas PJ. Metabolism of lithocholate in healthy man. III. Plasma disappearance of radioactivity after intravenous injection of labeled lithocholate and its derivatives. *Gastroenterology* 1975;69: pp. 77–82.

[258] Hofmann AF. Detoxification of lithocholic acid, a toxic bile acid: relevance to drug hepatotoxicity. *Drug Metab Rev* 2004;36: pp. 703–22.

[259] Kitada H, Miyata M, Nakamura T, Tozawa A, Honma W, Shimada M, Nagata K, et al. Protective role of hydroxysteroid sulfotransferase in lithocholic acid-induced liver toxicity. *J Biol Chem* 2003;278: pp. 17838–44.

[260] Francis GA, Fayard E, Picard F, Auwerx J. Nuclear receptors and the control of metabolism. *Annu Rev Physiol* 2003;65: pp. 261–311.

[261] Cai SY, Boyer JL. FXR: a target for cholestatic syndromes? *Expert Opin Ther Targets* 2006;10: pp. 409–21.

[262] Chiang JY. Bile acids: regulation of synthesis. *J Lipid Res* 2009;50: pp. 1955–66.

[263] Tu H, Okamoto AY, Shan B. FXR, a bile acid receptor and biological sensor. *Trends Cardiovasc Med* 2000;10: pp. 30–5.

[264] Moore JT, Goodwin B, Willson TM, Kliewer SA. Nuclear receptor regulation of genes involved in bile acid metabolism. *Crit Rev Eukaryot Gene Expr* 2002;12: pp. 119–35.

[265] Davis RA, Miyake JH, Hui TY, Spann NJ. Regulation of cholesterol-7alpha-hydroxylase: BAREly missing a SHP. *J Lipid Res* 2002;43: pp. 533–43.

[266] Fayard E, Schoonjans K, Auwerx J. Xol INXS: role of the liver X and the farnesol X receptors. *Curr Opin Lipidol* 2001;12: pp. 113–20.

[267] Beigneux A, Hofmann AF, Young SG. Human CYP7A1 deficiency: progress and enigmas. *J Clin Invest* 2002;110: pp. 29–31.

[268] Hofmann AF. Bile acids: trying to understand their chemistry and biology with the hope of helping patients. *Hepatology* 2009;49: pp. 1403–18.

[269] Hofmann AF. Biliary secretion and excretion in health and disease: current concepts. *Ann Hepatol* 2007;6: pp. 15–27.

[270] Kullak-Ublick GA, Stieger B, Meier PJ. Enterohepatic bile salt transporters in normal physiology and liver disease. *Gastroenterology* 2004;126: pp. 322–42.

[271] Fuchs M. Bile acid regulation of hepatic physiology: III. Regulation of bile acid synthesis: past progress and future challenges. *Am J Physiol Gastrointest Liver Physiol* 2003;284: pp. G551–7.

[272] Xu G, Li H, Pan LX, Shang Q, Honda A, Ananthanarayanan M, Erickson SK, et al. FXR-mediated down-regulation of CYP7A1 dominates LXRalpha in long-term cholesterol-fed NZW rabbits. *J Lipid Res* 2003;44: pp. 1956–62.

[273] Wang J, Einarsson C, Murphy C, Parini P, Bjorkhem I, Gafvels M, Eggertsen G. Studies on LXR- and FXR-mediated effects on cholesterol homeostasis in normal and cholic acid-depleted mice. *J Lipid Res* 2006;47: pp. 421–30.

[274] Wang L, Han Y, Kim CS, Lee YK, Moore DD. Resistance of SHP-null mice to bile acid-induced liver damage. *J Biol Chem* 2003;278: pp. 44475–81.

[275] Chiang JY, Kimmel R, Stroup D. Regulation of cholesterol 7alpha-hydroxylase gene (CYP7A1) transcription by the liver orphan receptor (LXRalpha). *Gene* 2001;262: pp. 257–65.

[276] Ellis E, Axelson M, Abrahamsson A, Eggertsen G, Thorne A, Nowak G, Ericzon BG, et al. Feedback regulation of bile acid synthesis in primary human hepatocytes: evidence that CDCA is the strongest inhibitor. *Hepatology* 2003;38: pp. 930–8.

[277] Eloranta JJ, Kullak-Ublick GA. Coordinate transcriptional regulation of bile acid homeostasis and drug metabolism. *Arch Biochem Biophys* 2005;433: pp. 397–412.

[278] Honda A, Salen G, Matsuzaki Y, Batta AK, Xu G, Hirayama T, Tint GS, et al. Disrupted coordinate regulation of farnesoid X receptor target genes in a patient with cerebrotendinous xanthomatosis. *J Lipid Res* 2005;46: pp. 287–96.

[279] Ito S, Fujimori T, Furuya A, Satoh J, Nabeshima Y. Impaired negative feedback suppression of bile acid synthesis in mice lacking betaKlotho. *J Clin Invest* 2005;115: pp. 2202–8.

[280] Kemper JK, Kim H, Miao J, Bhalla S, Bae Y. Role of an mSin3A-Swi/Snf chromatin remodeling complex in the feedback repression of bile acid biosynthesis by SHP. *Mol Cell Biol* 2004;24: pp. 7707–19.

[281] Kerr TA, Saeki S, Schneider M, Schaefer K, Berdy S, Redder T, Shan B, et al. Loss of nuclear receptor SHP impairs but does not eliminate negative feedback regulation of bile acid synthesis. *Dev Cell* 2002;2: pp. 713–20.

[282] Pullinger CR, Eng C, Salen G, Shefer S, Batta AK, Erickson SK, Verhagen A, et al. Human cholesterol 7alpha-hydroxylase (CYP7A1) deficiency has a hypercholesterolemic phenotype. *J Clin Invest* 2002;110: pp. 109–17.

[283] Li-Hawkins J, Gafvels M, Olin M, Lund EG, Andersson U, Schuster G, Bjorkhem I, et al. Cholic acid mediates negative feedback regulation of bile acid synthesis in mice. *J Clin Invest* 2002;110: pp. 1191–200.

[284] Chiang JY. Bile acid regulation of hepatic physiology: III. Bile acids and nuclear receptors. *Am J Physiol Gastrointest Liver Physiol* 2003;284: pp. G349–56.

[285] Chiang JY. Bile acid regulation of gene expression: roles of nuclear hormone receptors. *Endocr Rev* 2002;23: pp. 443–63.

[286] Chiang JY. Regulation of bile acid synthesis. *Front Biosci* 1998;3: pp. d176–93.

[287] Chiang JY. Regulation of bile acid synthesis: pathways, nuclear receptors, and mechanisms. *J Hepatol* 2004;40: pp. 539–51.

[288] Li T, Francl JM, Boehme S, Ochoa A, Zhang Y, Klaassen CD, Erickson SK, et al. Glucose and insulin induction of bile acid synthesis: mechanisms and implication in diabetes and obesity. *J Biol Chem* 2012;287: pp. 1861–73.

[289] Li T, Jahan A, Chiang JY. Bile acids and cytokines inhibit the human cholesterol 7 alpha-hydroxylase gene via the JNK/c-jun pathway in human liver cells. *Hepatology* 2006;43: pp. 1202–10.

[290] Li T, Kong X, Owsley E, Ellis E, Strom S, Chiang JY. Insulin regulation of cholesterol 7alpha-hydroxylase expression in human hepatocytes: roles of forkhead box O1 and sterol regulatory element-binding protein 1c. *J Biol Chem* 2006;281: pp. 28745–54.

[291] Li T, Chanda D, Zhang Y, Choi HS, Chiang JY. Glucose stimulates cholesterol 7alpha-hydroxylase gene transcription in human hepatocytes. *J Lipid Res* 2010;51: pp. 832–42.

[292] Li-Hawkins J, Lund EG, Turley SD, Russell DW. Disruption of the oxysterol 7alpha-hydroxylase gene in mice. *J Biol Chem* 2000;275: pp. 16536–42.

[293] Schwarz M, Lund EG, Setchell KD, Kayden HJ, Zerwekh JE, Bjorkhem I, Herz J, et al. Disruption of cholesterol 7alpha-hydroxylase gene in mice. II. Bile acid deficiency is overcome by induction of oxysterol 7alpha-hydroxylase. *J Biol Chem* 1996;271: pp. 18024–31.

[294] Hylemon PB, Stravitz RT, Vlahcevic ZR. Molecular genetics and regulation of bile acid biosynthesis. *Prog Liver Dis* 1994;12: pp. 99–120.

[295] Hylemon PB, Zhou H, Pandak WM, Ren S, Gil G, Dent P. Bile acids as regulatory molecules. *J Lipid Res* 2009;50: pp. 1509–20.

[296] Arrese M, Trauner M. Molecular aspects of bile formation and cholestasis. *Trends Mol Med* 2003;9: pp. 558–64.

[297] Bjorkhem I, Eggertsen G. Genes involved in initial steps of bile acid synthesis. *Curr Opin Lipidol* 2001;12: pp. 97–103.

[298] Goodwin B, Jones SA, Price RR, Watson MA, McKee DD, Moore LB, Galardi C, et al. A regulatory cascade of the nuclear receptors FXR, SHP-1, and LRH-1 represses bile acid biosynthesis. *Mol Cell* 2000;6: pp. 517–26.

[299] Wang H, Chen J, Hollister K, Sowers LC, Forman BM. Endogenous bile acids are ligands for the nuclear receptor FXR/BAR. *Mol Cell* 1999;3: pp. 543–53.

[300] Lu TT, Makishima M, Repa JJ, Schoonjans K, Kerr TA, Auwerx J, Mangelsdorf DJ.

Molecular basis for feedback regulation of bile acid synthesis by nuclear receptors. *Mol Cell* 2000;6: pp. 507–15.

[301] Downes M, Verdecia MA, Roecker AJ, Hughes R, Hogenesch JB, Kast-Woelbern HR, Bowman ME, et al. A chemical, genetic, and structural analysis of the nuclear bile acid receptor FXR. *Mol Cell* 2003;11: pp. 1079–92.

[302] Eloranta JJ, Meier PJ, Kullak-Ublick GA. Coordinate transcriptional regulation of transport and metabolism. *Methods Enzymol* 2005;400: pp. 511–30.

[303] Wang L, Lee YK, Bundman D, Han Y, Thevananther S, Kim CS, Chua SS, et al. Redundant pathways for negative feedback regulation of bile acid production. *Dev Cell* 2002;2: pp. 721–31.

[304] Gupta S, Pandak WM, Hylemon PB. LXR alpha is the dominant regulator of CYP7A1 transcription. *Biochem Biophys Res Commun* 2002;293: pp. 338–43.

[305] Miao J, Choi SE, Seok SM, Yang L, Zuercher WJ, Xu Y, Willson TM, et al. Ligand-dependent regulation of the activity of the orphan nuclear receptor, small heterodimer partner (SHP), in the repression of bile acid biosynthetic CYP7A1 and CYP8B1 Genes. *Mol Endocrinol* 2011;25: pp. 1159–69.

[306] Boulias K, Katrakili N, Bamberg K, Underhill P, Greenfield A, Talianidis I. Regulation of hepatic metabolic pathways by the orphan nuclear receptor SHP. *EMBO J* 2005;24: pp. 2624–33.

[307] Repa JJ, Mangelsdorf DJ. Nuclear receptor regulation of cholesterol and bile acid metabolism. *Curr Opin Biotechnol* 1999;10: pp. 557–63.

[308] Song KH, Li T, Owsley E, Strom S, Chiang JY. Bile acids activate fibroblast growth factor 19 signaling in human hepatocytes to inhibit cholesterol 7alpha-hydroxylase gene expression. *Hepatology* 2009;49: pp. 297–305.

[309] Miyata M, Takamatsu Y, Kuribayashi H, Yamazoe Y. Administration of ampicillin elevates hepatic primary bile acid synthesis through suppression of ileal fibroblast growth factor 15 expression. *J Pharmacol Exp Ther* 2009;331: pp. 1079–85.

[310] Schaap FG, van der Gaag NA, Gouma DJ, Jansen PL. High expression of the bile salt-homeostatic hormone fibroblast growth factor 19 in the liver of patients with extrahepatic cholestasis. *Hepatology* 2009;49: pp. 1228–35.

[311] Stroeve JH, Brufau G, Stellaard F, Gonzalez FJ, Staels B, Kuipers F. Intestinal FXR-mediated FGF15 production contributes to diurnal control of hepatic bile acid synthesis in mice. *Lab Invest* 2010;90: pp. 1457–67.

[312] Inagaki T, Choi M, Moschetta A, Peng L, Cummins CL, McDonald JG, Luo G, et al. Fibroblast growth factor 15 functions as an enterohepatic signal to regulate bile acid homeostasis. *Cell Metab* 2005;2: pp. 217–25.

[313] Jones S. Mini-review: endocrine actions of fibroblast growth factor 19. *Mol Pharm* 2008;5: pp. 42–8.

[314] Gutierrez A, Ratliff EP, Andres AM, Huang X, McKeehan WL, Davis RA. Bile acids decrease hepatic paraoxonase 1 expression and plasma high-density lipoprotein levels via FXR-mediated signaling of FGFR4. *Arterioscler Thromb Vasc Biol* 2006;26: pp. 301–6.

[315] Sinha J, Chen F, Miloh T, Burns RC, Yu Z, Shneider BL. beta-Klotho and FGF-15/19 inhibit the apical sodium-dependent bile acid transporter in enterocytes and cholangiocytes. *Am J Physiol Gastrointest Liver Physiol* 2008;295: pp. G996–1003.

[316] del Castillo-Olivares A, Campos JA, Pandak WM, Gil G. The role of alpha1-fetoprotein transcription factor/LRH-1 in bile acid biosynthesis: a known nuclear receptor activator that can act as a suppressor of bile acid biosynthesis. *J Biol Chem* 2004;279: pp. 16813–21.

[317] Chen W, Chiang JY. Regulation of human sterol 27-hydroxylase gene (CYP27A1) by bile acids and hepatocyte nuclear factor 4alpha (HNF4alpha). *Gene* 2003;313: pp. 71–82.

[318] Chiang JY. Hepatocyte nuclear factor 4alpha regulation of bile acid and drug metabolism. *Expert Opin Drug Metab Toxicol* 2009;5: pp. 137–47.

[319] Hayhurst GP, Lee YH, Lambert G, Ward JM, Gonzalez FJ. Hepatocyte nuclear factor 4alpha (nuclear receptor 2A1) is essential for maintenance of hepatic gene expression and lipid homeostasis. *Mol Cell Biol* 2001;21: pp. 1393–403.

[320] Fayard E, Auwerx J, Schoonjans K. LRH-1: an orphan nuclear receptor involved in development, metabolism and steroidogenesis. *Trends Cell Biol* 2004;14: pp. 250–60.

[321] Kir S, Kliewer SA, Mangelsdorf DJ. Roles of FGF19 in liver metabolism. *Cold Spring Harb Symp Quant Biol* 2011.

[322] Triantis V, Saeland E, Bijl N, Oude-Elferink RP, Jansen PL. Glycosylation of fibroblast growth factor receptor 4 is a key regulator of fibroblast growth factor 19-mediated down-regulation of cytochrome P450 7A1. *Hepatology* 2010;52: pp. 656–66.

[323] Lundasen T, Galman C, Angelin B, Rudling M. Circulating intestinal fibroblast growth factor 19 has a pronounced diurnal variation and modulates hepatic bile acid synthesis in man. *J Intern Med* 2006;260: pp. 530–6.

[324] Durovcova V, Marek J, Hana V, Matoulek M, Zikan V, Haluzikova D, Kavalkova P, et al. Plasma concentrations of adipocyte fatty acid binding protein in patients with Cushing's syndrome. *Physiol Res* 2010;59: pp. 963–71.

[325] Dostalova I, Kavalkova P, Haluzikova D, Lacinova Z, Mraz M, Papezova H, Haluzik M. Plasma concentrations of fibroblast growth factors 19 and 21 in patients with anorexia nervosa. *J Clin Endocrinol Metab* 2008;93: pp. 3627–32.

[326] Holt JA, Luo G, Billin AN, Bisi J, McNeill YY, Kozarsky KF, Donahee M, et al. Definition

of a novel growth factor-dependent signal cascade for the suppression of bile acid biosynthesis. *Genes Dev* 2003;17: pp. 1581–91.

[327] Kim I, Ahn SH, Inagaki T, Choi M, Ito S, Guo GL, Kliewer SA, et al. Differential regulation of bile acid homeostasis by the farnesoid X receptor in liver and intestine. *J Lipid Res* 2007;48: pp. 2664–72.

[328] Chatterjee B, Echchgadda I, Song CS. Vitamin D receptor regulation of the steroid/bile acid sulfotransferase SULT2A1. *Methods Enzymol* 2005;400: pp. 165–91.

[329] Chawla A, Repa JJ, Evans RM, Mangelsdorf DJ. Nuclear receptors and lipid physiology: opening the X-files. *Science* 2001;294: pp. 1866–70.

[330] Adachi R, Honma Y, Masuno H, Kawana K, Shimomura I, Yamada S, Makishima M. Selective activation of vitamin D receptor by lithocholic acid acetate, a bile acid derivative. *J Lipid Res* 2005;46: pp. 46–57.

[331] Han S, Li T, Ellis E, Strom S, Chiang JY. A novel bile acid-activated vitamin D receptor signaling in human hepatocytes. *Mol Endocrinol* 2010;24: pp. 1151–64.

[332] Makishima M, Lu TT, Xie W, Whitfield GK, Domoto H, Evans RM, Haussler MR, et al. Vitamin D receptor as an intestinal bile acid sensor. *Science* 2002;296: pp. 1313–6.

[333] Matsubara T, Yoshinari K, Aoyama K, Sugawara M, Sekiya Y, Nagata K, Yamazoe Y. Role of vitamin D receptor in the lithocholic acid-mediated CYP3A induction in vitro and in vivo. *Drug Metab Dispos* 2008;36: pp. 2058–63.

[334] Ogura M, Nishida S, Ishizawa M, Sakurai K, Shimizu M, Matsuo S, Amano S, et al. Vitamin D3 modulates the expression of bile acid regulatory genes and represses inflammation in bile duct-ligated mice. *J Pharmacol Exp Ther* 2009;328: pp. 564–70.

[335] Jiang W, Miyamoto T, Kakizawa T, Nishio SI, Oiwa A, Takeda T, Suzuki S, et al. Inhibition of LXRalpha signaling by vitamin D receptor: possible role of VDR in bile acid synthesis. *Biochem Biophys Res Commun* 2006;351: pp. 176–84.

[336] Moore DD, Kato S, Xie W, Mangelsdorf DJ, Schmidt DR, Xiao R, Kliewer SA. International Union of Pharmacology. LXII. The NR1H and NR1I receptors: constitutive androstane receptor, pregnene X receptor, farnesoid X receptor alpha, farnesoid X receptor beta, liver X receptor alpha, liver X receptor beta, and vitamin D receptor. *Pharmacol Rev* 2006;58: pp. 742–59.

[337] Schmidt DR, Holmstrom SR, Fon Tacer K, Bookout AL, Kliewer SA, Mangelsdorf DJ. Regulation of bile acid synthesis by fat-soluble vitamins A and D. *J Biol Chem* 2010;285: pp. 14486–94.

[338] Allen K, Kim ND, Moon JO, Copple BL. Upregulation of early growth response factor-1 by bile acids requires mitogen-activated protein kinase signaling. *Toxicol Appl Pharmacol* 2010;243: pp. 63–7.

[339] Rao YP, Studer EJ, Stravitz RT, Gupta S, Qiao L, Dent P, Hylemon PB. Activation of the Raf-1/MEK/ERK cascade by bile acids occurs via the epidermal growth factor receptor in primary rat hepatocytes. *Hepatology* 2002;35: pp. 307–14.

[340] Dent P, Han SI, Mitchell C, Studer E, Yacoub A, Grandis J, Grant S, et al. Inhibition of insulin/IGF-1 receptor signaling enhances bile acid toxicity in primary hepatocytes. *Biochem Pharmacol* 2005;70: pp. 1685–96.

[341] Huang W, Ma K, Zhang J, Qatanani M, Cuvillier J, Liu J, Dong B, et al. Nuclear receptor-dependent bile acid signaling is required for normal liver regeneration. *Science* 2006;312: pp. 233–6.

[342] Li T, Ma H, Chiang JY. TGFbeta1, TNFalpha, and insulin signaling crosstalk in regulation of the rat cholesterol 7alpha-hydroxylase gene expression. *J Lipid Res* 2008;49: pp. 1981–9.

[343] Gadaleta RM, Oldenburg B, Willemsen EC, Spit M, Murzilli S, Salvatore L, Klomp LW, et al. Activation of bile salt nuclear receptor FXR is repressed by pro-inflammatory cytokines activating NF-kappaB signaling in the intestine. *Biochim Biophys Acta* 2011;1812: pp. 851–8.

[344] Lickteig AJ, Slitt AL, Arkan MC, Karin M, Cherrington NJ. Differential regulation of hepatic transporters in the absence of tumor necrosis factor-alpha, interleukin-1beta, interleukin-6, and nuclear factor-kappaB in two models of cholestasis. *Drug Metab Dispos* 2007;35: pp. 402–9.

[345] Qiao L, Han SI, Fang Y, Park JS, Gupta S, Gilfor D, Amorino G, et al. Bile acid regulation of C/EBPbeta, CREB, and c-Jun function, via the extracellular signal-regulated kinase and c-Jun NH2-terminal kinase pathways, modulates the apoptotic response of hepatocytes. *Mol Cell Biol* 2003;23: pp. 3052–66.

[346] Li D, Zimmerman TL, Thevananther S, Lee HY, Kurie JM, Karpen SJ. Interleukin-1 beta-mediated suppression of RXR:RAR transactivation of the Ntcp promoter is JNK-dependent. *J Biol Chem* 2002;277: pp. 31416–22.

[347] Gupta S, Natarajan R, Payne SG, Studer EJ, Spiegel S, Dent P, Hylemon PB. Deoxycholic acid activates the c-Jun N-terminal kinase pathway via FAS receptor activation in primary hepatocytes. Role of acidic sphingomyelinase-mediated ceramide generation in FAS receptor activation. *J Biol Chem* 2004;279: pp. 5821–8.

[348] Gupta S, Stravitz RT, Dent P, Hylemon PB. Down-regulation of cholesterol 7alpha-hydroxylase (CYP7A1) gene expression by bile acids in primary rat hepatocytes is mediated by the c-Jun N-terminal kinase pathway. *J Biol Chem* 2001;276: pp. 15816–22.

[349] Zimmerman TL, Thevananther S, Ghose R, Burns AR, Karpen SJ. Nuclear export of retinoid X receptor alpha in response to interleukin-1beta-mediated cell signaling: roles for JNK and SER260. *J Biol Chem* 2006;281: pp. 15434–40.

[350] Yang JI, Yoon JH, Myung SJ, Gwak GY, Kim W, Chung GE, Lee SH, et al. Bile acid-induced TGR5-dependent c-Jun-N terminal kinase activation leads to enhanced caspase 8 activation in hepatocytes. *Biochem Biophys Res Commun* 2007;361: pp. 156–61.

[351] Dawson PA: Bile Secretion and the Enterohepatic Circulation. In: Feldman M, Friedman LS, Brandt L, eds. Sleisenger and Fordtran's Gastrointestinal and Liver Disease. 9 ed. Philadelphia: Elsevier Saunders, 2010; pp. 1075–88.

[352] Hofmann AF: Bile Acids and the Enterohepatic Circulation. In: Arias IM, Alter HJ, Boyer JL, Cohen DE, Fausto N, Shafritz DA, Wolkoff AW, eds. The Liver: Biology and Pathobiology. 5 ed. West Sussex: Wiley-Blackwell, 2009; pp. 290–304.

[353] Hofmann AF. The enterohepatic circulation of bile acids in mammals: form and functions. *Front Biosci* 2009;14: pp. 2584–98.

[354] Dawson PA, Haywood J, Craddock AL, Wilson M, Tietjen M, Kluckman K, Maeda N, et al. Targeted deletion of the ileal bile acid transporter eliminates enterohepatic cycling of bile acids in mice. *J Biol Chem* 2003;278: pp. 33920–7.

[355] Portincasa P, Di Ciaula A, Wang HH, Palasciano G, van Erpecum KJ, Moschetta A, Wang DQ. Coordinate regulation of gallbladder motor function in the gut-liver axis. *Hepatology* 2008;47: pp. 2112–26.

[356] Carey MC, Small DM, Bliss CM. Lipid digestion and absorption. *Annu Rev Physiol* 1983;45: pp. 651–77.

[357] Staggers JE, Hernell O, Stafford RJ, Carey MC. Physical–chemical behavior of dietary and biliary lipids during intestinal digestion and absorption. 1. Phase behavior and aggregation states of model lipid systems patterned after aqueous duodenal contents of healthy adult human beings. *Biochemistry* 1990;29: pp. 2028–40.

[358] Hernell O, Staggers JE, Carey MC. Physical–chemical behavior of dietary and biliary lipids during intestinal digestion and absorption. 2. Phase analysis and aggregation states of luminal lipids during duodenal fat digestion in healthy adult human beings. *Biochemistry* 1990;29: pp. 2041–56.

[359] Xu G, Salen G: Nuclear Receptors Regulate Bile Acid Synthesis. In: Arias IM, Alter HJ, Boyer JL, Cohen DE, Fausto N, Shafritz DA, Wolkoff AW, eds. The Liver: Biology and Pathobiology. 5 ed. West Sussex: Wiley-Blackwell, 2009; pp. 323–48.

[360] Mok HY, Von Bergmann K, Grundy SM. Regulation of pool size of bile acids in man. *Gastroenterology* 1977;73: pp. 684–90.

[361] Mok HY, von Bergmann K, Grundy SM. Effects of continuous and intermittent feeding on biliary lipid outputs in man: application for measurements of intestinal absorption of cholesterol and bile acids. *J Lipid Res* 1979;20: pp. 389–98.

[362] Mok HY, von Bergmann K, Crouse JR, Grundy SM. Bilary lipid metabolism in obesity.

Effects of bile acid feeding before and during weight reduction. *Gastroenterology* 1979;76: pp. 556–67.

[363] Eriksson S. Biliary excretion of bile acids and cholesterol in bile fistula rats; bile acids and steroids. *Proc Soc Exp Biol Med* 1957;94: pp. 578–82.

[364] Wang DQ, Lammert F, Paigen B, Carey MC. Phenotypic characterization of lith genes that determine susceptibility to cholesterol cholelithiasis in inbred mice. Pathophysiology Of biliary lipid secretion. *J Lipid Res* 1999;40: pp. 2066–79.

[365] Dowling RH, Mack E, Picott J, Berger J, Small DM. Experimental model for the study of the enterohepatic circulation of bile in rhesus monkeys. *J Lab Clin Med* 1968;72: pp. 169–76.

[366] Dowling RH, Mack E, Small DM. Biliary lipid secretion and bile composition after acute and chronic interruption of the enterohepatic circulation in the Rhesus monkey. IV. Primate biliary physiology. *J Clin Invest* 1971;50: pp. 1917–26.

[367] Dowling RH, Mack E, Small DM. Effects of controlled interruption of the enterohepatic circulation of bile salts by biliary diversion and by ileal resection on bile salt secretion, synthesis, and pool size in the rhesus monkey. *J Clin Invest* 1970;49: pp. 232–42.

[368] Suchy FJ, Ananthanarayanan M. Bile salt excretory pump: biology and pathobiology. *J Pediatr Gastroenterol Nutr* 2006;43 Suppl 1: pp. S10–6.

[369] Trauner M, Boyer JL. Bile salt transporters: molecular characterization, function, and regulation. *Physiol Rev* 2003;83: pp. 633–71.

[370] Gerloff T, Stieger B, Hagenbuch B, Madon J, Landmann L, Roth J, Hofmann AF, et al. The sister of P-glycoprotein represents the canalicular bile salt export pump of mammalian liver. *J Biol Chem* 1998;273: pp. 10046–50.

[371] Wang R, Salem M, Yousef IM, Tuchweber B, Lam P, Childs SJ, Helgason CD, et al. Targeted inactivation of sister of P-glycoprotein gene (spgp) in mice results in nonprogressive but persistent intrahepatic cholestasis. *Proc Natl Acad Sci U S A* 2001;98: pp. 2011–6.

[372] Wang R, Lam P, Liu L, Forrest D, Yousef IM, Mignault D, Phillips MJ, et al. Severe cholestasis induced by cholic acid feeding in knockout mice of sister of P-glycoprotein. *Hepatology* 2003;38: pp. 1489–99.

[373] Oude Elferink RP, Paulusma CC: The Function of the Canalicular Membrane in Bile Formation and Secretion. In: Arias IM, Alter HJ, Boyer JL, Cohen DE, Fausto N, Shafritz DA, Wolkoff AW, eds. The Liver: Biology and Pathobiology. 5 ed. West Sussex: Wiley-Blackwell, 2009; pp. 339–48.

[374] Hagenbuch B, Meier PJ. The superfamily of organic anion transporting polypeptides. *Biochim Biophys Acta* 2003;1609: pp. 1–18.

[375] Hagenbuch B, Meier PJ. Sinusoidal (basolateral) bile salt uptake systems of hepatocytes. *Semin Liver Dis* 1996;16: pp. 129–36.

[376] Hagenbuch B, Meier PJ. Molecular cloning, chromosomal localization, and functional characterization of a human liver Na+/bile acid cotransporter. *J Clin Invest* 1994;93: pp. 1326–31.

[377] Bossuyt X, Muller M, Meier PJ. Multispecific amphipathic substrate transport by an organic anion transporter of human liver. *J Hepatol* 1996;25: pp. 733–8.

[378] Boyer JL, Hagenbuch B, Ananthanarayanan M, Suchy F, Stieger B, Meier PJ. Phylogenic and ontogenic expression of hepatocellular bile acid transport. *Proc Natl Acad Sci U S A* 1993;90: pp. 435–8.

[379] Ballatori N, Christian WV, Lee JY, Dawson PA, Soroka CJ, Boyer JL, Madejczyk MS, et al. OSTalpha-OSTbeta: a major basolateral bile acid and steroid transporter in human intestinal, renal, and biliary epithelia. *Hepatology* 2005;42: pp. 1270–9.

[380] Ballatori N, Li N, Fang F, Boyer JL, Christian WV, Hammond CL. OST alpha-OST beta: a key membrane transporter of bile acids and conjugated steroids. *Front Biosci* 2009;14: pp. 2829–44.

[381] Soroka CJ, Ballatori N, Boyer JL. Organic solute transporter, OSTalpha-OSTbeta: its role in bile acid transport and cholestasis. *Semin Liver Dis* 2010;30: pp. 178–85.

[382] Cattori V, van Montfoort JE, Stieger B, Landmann L, Meijer DK, Winterhalter KH, Meier PJ, et al. Localization of organic anion transporting polypeptide 4 (Oatp4) in rat liver and comparison of its substrate specificity with Oatp1, Oatp2 and Oatp3. *Pflugers Arch* 2001;443: pp. 188–95.

[383] Oude Elferink RP, Paulusma CC, Groen AK. Hepatocanalicular transport defects: pathophysiologic mechanisms of rare diseases. *Gastroenterology* 2006;130: pp. 908–25.

[384] Stapelbroek JM, van Erpecum KJ, Klomp LW, Houwen RH. Liver disease associated with canalicular transport defects: current and future therapies. *J Hepatol* 2010;52: pp. 258–71.

[385] Scheimann AO, Strautnieks SS, Knisely AS, Byrne JA, Thompson RJ, Finegold MJ. Mutations in bile salt export pump (ABCB11) in two children with progressive familial intrahepatic cholestasis and cholangiocarcinoma. *J Pediatr* 2007;150: pp. 556–9.

[386] Strautnieks SS, Byrne JA, Pawlikowska L, Cebecauerova D, Rayner A, Dutton L, Meier Y, et al. Severe bile salt export pump deficiency: 82 different ABCB11 mutations in 109 families. *Gastroenterology* 2008;134: pp. 1203–14.

[387] Lam P, Wang R, Ling V. Bile acid transport in sister of P-glycoprotein (ABCB11) knockout mice. *Biochemistry* 2005;44: pp. 12598–605.

[388] Schalm SW, LaRusso NF, Hofmann AF, Hoffman NE, van Berge-Henegouwen GP, Korman MG. Diurnal serum levels of primary conjugated bile acids. Assessment by specific radioimmunoassays for conjugates of cholic and chenodeoxycholic acid. *Gut* 1978;19: pp. 1006–14.

[389] Ponz De Leon M, Murphy GM, Dowling RH. Physiological factors influencing serum bile acid levels. *Gut* 1978;19: pp. 32–9.

[390] LaRusso NF, Korman MG, Hoffman NE, Hofmann AF. Dynamics of the enterohepatic circulation of bile acids. Postprandial serum concentrations of conjugates of cholic acid in health, cholecystectomized patients, and patients with bile acid malabsorption. *N Engl J Med* 1974;291: pp. 689–92.

[391] Choi M, Moschetta A, Bookout AL, Peng L, Umetani M, Holmstrom SR, Suino-Powell K, et al. Identification of a hormonal basis for gallbladder filling. *Nat Med* 2006; 12: pp. 1253–5.

[392] Fisher RS, Rock E, Malmud LS. Gallbladder emptying response to sham feeding in humans. *Gastroenterology* 1986;90: pp. 1854–7.

[393] Shaffer EA, McOrmond P, Duggan H. Quantitative cholescintigraphy: assessment of gallbladder filling and emptying and duodenogastric reflux. *Gastroenterology* 1980;79: pp. 899–906.

[394] Portincasa P, Di Ciaula A, vanBerge-Henegouwen GP. Smooth muscle function and dysfunction in gallbladder disease. *Curr Gastroenterol Rep* 2004;6: pp. 151–62.

[395] Lawson M, Everson GT, Klingensmith W, Kern F, Jr. Coordination of gastric and gallbladder emptying after ingestion of a regular meal. *Gastroenterology* 1983;85: pp. 866–70.

[396] Lanzini A, Jazrawi RP, Northfield TC. Simultaneous quantitative measurements of absolute gallbladder storage and emptying during fasting and eating in humans. *Gastroenterology* 1987;92: pp. 852–861.

[397] Carey MC. Pathogenesis of gallstones. *Am J Surg* 1993;165: pp. 410–9.

[398] Hagey LR, Schteingart CD, Ton-Nu HT, Hofmann AF. Biliary bile acids of fruit pigeons and doves (Columbiformes): presence of 1-beta-hydroxychenodeoxycholic acid and conjugation with glycine as well as taurine. *J Lipid Res* 1994;35: pp. 2041–8.

[399] Hagey LR, Vidal N, Hofmann AF, Krasowski MD. Complex Evolution of Bile Salts in Birds. *Auk* 2010;127: pp. 820–31.

[400] Hagey LR, Vidal N, Hofmann AF, Krasowski MD. Evolutionary diversity of bile salts in reptiles and mammals, including analysis of ancient human and extinct giant ground sloth coprolites. *BMC Evol Biol* 2010;10: p. 133.

[401] Almond HR, Vlahcevic ZR, Bell CC, Jr., Gregory DH, Swell L. Bile acid pools, kinetics and biliary lipid composition before and after cholecystectomy. *N Engl J Med* 1973;289: pp. 1213–6.

[402] Bell CC, Jr., McCormick WC, 3rd, Gregory DH, Law DH, Vlahcevic ZR, Swell L. Relationship of bile acid pool size to the formation of lithogenous bile in male Indians of the Southwest. *Surg Gynecol Obstet* 1972;134: pp. 473–8.

[403] Kimball A, Pertsemlidis D, Panveliwalla D. Composition of biliary lipids and kinetics of bile acids after cholecystectomy in man. *Am J Dig Dis* 1976;21: pp. 776–81.

[404] Low-Beer TS, Pomare EW. Regulation of bile salt pool size in man. *Br Med J* 1973;2: pp. 338–40.

[405] Shneider BL, Dawson PA, Christie DM, Hardikar W, Wong MH, Suchy FJ. Cloning and molecular characterization of the ontogeny of a rat ileal sodium-dependent bile acid transporter. *J Clin Invest* 1995;95: pp. 745–54.

[406] Oelkers P, Kirby LC, Heubi JE, Dawson PA. Primary bile acid malabsorption caused by mutations in the ileal sodium-dependent bile acid transporter gene (SLC10A2). *J Clin Invest* 1997;99: pp. 1880–7.

[407] Montagnani M, Abrahamsson A, Galman C, Eggertsen G, Marschall HU, Ravaioli E, Einarsson C, et al. Analysis of ileal sodium/bile acid cotransporter and related nuclear receptor genes in a family with multiple cases of idiopathic bile acid malabsorption. *World J Gastroenterol* 2006;12: pp. 7710–4.

[408] Mottino AD, Hoffman T, Dawson PA, Luquita MG, Monti JA, Sanchez Pozzi EJ, Catania VA, et al. Increased expression of ileal apical sodium-dependent bile acid transporter in postpartum rats. *Am J Physiol Gastrointest Liver Physiol* 2002;282: pp. G41–50.

[409] Dawson PA, Hubbert M, Haywood J, Craddock AL, Zerangue N, Christian WV, Ballatori N. The heteromeric organic solute transporter alpha-beta, Ostalpha-Ostbeta, is an ileal basolateral bile acid transporter. *J Biol Chem* 2005;280: pp. 6960–8.

[410] Dawson PA, Hubbert ML, Rao A. Getting the mOST from OST: role of organic solute transporter, OSTalpha-OSTbeta, in bile acid and steroid metabolism. *Biochim Biophys Acta* 2010;1801: pp. 994–1004.

[411] Rao A, Haywood J, Craddock AL, Belinsky MG, Kruh GD, Dawson PA. The organic solute transporter alpha-beta, Ostalpha-Ostbeta, is essential for intestinal bile acid transport and homeostasis. *Proc Natl Acad Sci U S A* 2008;105: pp. 3891–6.

[412] Frankenberg T, Rao A, Chen F, Haywood J, Shneider BL, Dawson PA. Regulation of the mouse organic solute transporter alpha-beta, Ostalpha-Ostbeta, by bile acids. *Am J Physiol Gastrointest Liver Physiol* 2006;290: pp. G912–22.

[413] Belinsky MG, Dawson PA, Shchaveleva I, Bain LJ, Wang R, Ling V, Chen ZS, et al. Analysis of the in vivo functions of Mrp3. *Mol Pharmacol* 2005;68: pp. 160–8.

[414] Ogawa K, Suzuki H, Hirohashi T, Ishikawa T, Meier PJ, Hirose K, Akizawa T, et al. Characterization of inducible nature of MRP3 in rat liver. *Am J Physiol Gastrointest Liver Physiol* 2000;278: pp. G438–46.

[415] Kullak-Ublick GA, Becker MB. Regulation of drug and bile salt transporters in liver and intestine. *Drug Metab Rev* 2003;35: pp. 305–17.

[416] Kullak-Ublick GA, Stieger B, Hagenbuch B, Meier PJ. Hepatic transport of bile salts. *Semin Liver Dis* 2000;20: pp. 273–92.

[417] Wong MH, Rao PN, Pettenati MJ, Dawson PA. Localization of the ileal sodium-bile acid cotransporter gene (SLC10A2) to human chromosome 13q33. *Genomics* 1996;33: pp. 538–40.

[418] Jung D, Fried M, Kullak-Ublick GA. Human apical sodium-dependent bile salt transporter gene (SLC10A2) is regulated by the peroxisome proliferator-activated receptor alpha. *J Biol Chem* 2002;277: pp. 30559–66.

[419] Lammert F, Paigen B, Carey MC. Localization of the ileal sodium-bile salt cotransporter gene (Slc10a2) to mouse chromosome 8. *Mamm Genome* 1998;9: pp. 173–4.

[420] Love MW, Craddock AL, Angelin B, Brunzell JD, Duane WC, Dawson PA. Analysis of the ileal bile acid transporter gene, SLC10A2, in subjects with familial hypertriglyceridemia. *Arterioscler Thromb Vasc Biol* 2001;21: pp. 2039–45.

[421] Montagnani M, Love MW, Rossel P, Dawson PA, Qvist P. Absence of dysfunctional ileal sodium-bile acid cotransporter gene mutations in patients with adult-onset idiopathic bile acid malabsorption. *Scand J Gastroenterol* 2001;36: pp. 1077–80.

[422] Wang DQ, Zhang L, Wang HH. High cholesterol absorption efficiency and rapid biliary secretion of chylomicron remnant cholesterol enhance cholelithogenesis in gallstone-susceptible mice. *Biochim Biophys Acta* 2005;1733: pp. 90–9.

[423] Hofmann AF, Hagey LR. Bile acids: chemistry, pathochemistry, biology, pathobiology, and therapeutics. *Cell Mol Life Sci* 2008;65: pp. 2461–83.

[424] Somjen GJ, Marikovsky Y, Lelkes P, Gilat T. Cholesterol-phospholipid vesicles in human bile: an ultrastructural study. *Biochim Biophys Acta* 1986;879: pp. 14–21.

[425] Mazer NA, Carey MC, Kwasnick RF, Benedek GB. Quasielastic light scattering studies of aqueous biliary lipid systems. Size, shape, and thermodynamics of bile salt micelles. *Biochemistry* 1979;18: pp. 3064–75.

[426] Mazer NA, Benedek GB, Carey MC. Quasielastic light-scattering studies of aqueous biliary lipid systems. Mixed micelle formation in bile salt-lecithin solutions. *Biochemistry* 1980;19: pp. 601–15.

[427] Mazer NA, Schurtenberg P, Carey MC, Preisig R, Weigand K, Kanzig W. Quasi-elastic light scattering studies of native hepatic bile from the dog: comparison with aggregative behavior of model biliary lipid systems. *Biochemistry* 1984;23: pp. 1994–2005.

[428] Somjen GJ, Gilat T. Contribution of vesicular and micellar carriers to cholesterol transport in human bile. *J Lipid Res* 1985;26: pp. 699–704.

[429] Somjen GJ, Gilat T. A non-micellar mode of cholesterol transport in human bile. *FEBS Lett* 1983;156: pp. 265–8.

[430] Crawford JM, Mockel GM, Crawford AR, Hagen SJ, Hatch VC, Barnes S, Godleski JJ, et al. Imaging biliary lipid secretion in the rat: ultrastructural evidence for vesiculation of the hepatocyte canalicular membrane. *J Lipid Res* 1995;36: pp. 2147–63.

[431] Crawford JM. Role of vesicle-mediated transport pathways in hepatocellular bile secretion. *Semin Liver Dis* 1996;16: pp. 169–89.

[432] Crawford AR, Smith AJ, Hatch VC, Oude Elferink RP, Borst P, Crawford JM. Hepatic secretion of phospholipid vesicles in the mouse critically depends on mdr2 or MDR3 P-glycoprotein expression. Visualization by electron microscopy. *J Clin Invest* 1997;100: pp. 2562–7.

[433] Cohen DE, Fisch MR, Carey MC. Principles of laser light-scattering spectroscopy: applications to the physicochemical study of model and native biles. *Hepatology* 1990;12:113S–21S; discussion 121S–2S.

[434] Mazer NA, Carey MC. Quasi-elastic light-scattering studies of aqueous biliary lipid systems. Cholesterol solubilization and precipitation in model bile solutions. *Biochemistry* 1983;22: pp. 426–42.

[435] Ulloa N, Garrido J, Nervi F. Ultracentrifugal isolation of vesicular carriers of biliary cholesterol in native human and rat bile. *Hepatology* 1987;7: pp. 235–44.

[436] Cohen DE. Hepatocellular transport and secretion of biliary lipids. *Curr Opin Lipidol* 1999;10: pp. 295–302.

[437] Cohen DE, Thurston GM, Chamberlin RA, Benedek GB, Carey MC. Laser light scattering evidence for a common wormlike growth structure of mixed micelles in bile salt- and straight-chain detergent-phosphatidylcholine aqueous systems: relevance to the micellar structure of bile. *Biochemistry* 1998;37: pp. 14798–814.

[438] Elferink RO, Groen AK. Genetic defects in hepatobiliary transport. *Biochim Biophys Acta* 2002;1586: pp. 129–45.

[439] Claudel T, Zollner G, Wagner M, Trauner M. Role of nuclear receptors for bile acid metabolism, bile secretion, cholestasis, and gallstone disease. *Biochim Biophys Acta* 2011;1812: pp. 867–78.

[440] Wagner M, Zollner G, Trauner M. Nuclear receptor regulation of the adaptive response of bile acid transporters in cholestasis. *Semin Liver Dis* 2010;30: pp. 160–77.

[441] Bhattacharyya AK, Connor WE. Beta-sitosterolemia and xanthomatosis. A newly described lipid storage disease in two sisters. *J Clin Invest* 1974;53: pp. 1033–43.

[442] Berge KE. Sitosterolemia: a gateway to new knowledge about cholesterol metabolism. *Ann Med* 2003;35: pp. 502–11.

[443] Salen G, Patel S, Batta AK. Sitosterolemia. *Cardiovasc Drug Rev* 2002;20: pp. 255–70.

[444] Salen G, Shefer S, Nguyen L, Ness GC, Tint GS, Batta AK. Sitosterolemia. *Subcell Biochem* 1997;28: pp. 453–76.

[445] Salen G, Shefer S, Nguyen L, Ness GC, Tint GS, Shore V. Sitosterolemia. *J Lipid Res* 1992;33: pp. 945–55.

[446] Salen G, Shore V, Tint GS, Forte T, Shefer S, Horak I, Horak E, et al. Increased sitosterol absorption, decreased removal, and expanded body pools compensate for reduced cholesterol synthesis in sitosterolemia with xanthomatosis. *J Lipid Res* 1989;30: pp. 1319–30.

[447] Lu K, Lee MH, Yu H, Zhou Y, Sandell SA, Salen G, Patel SB. Molecular cloning, genomic organization, genetic variations, and characterization of murine sterolin genes Abcg5 and Abcg8. *J Lipid Res* 2002;43: pp. 565–78.

[448] Lu K, Lee MH, Hazard S, Brooks-Wilson A, Hidaka H, Kojima H, Ose L, et al. Two genes that map to the STSL locus cause sitosterolemia: genomic structure and spectrum of mutations involving sterolin-1 and sterolin-2, encoded by ABCG5 and ABCG8, respectively. *Am J Hum Genet* 2001;69: pp. 278–90.

[449] Nguyen LB, Shefer S, Salen G, Ness GC, Tint GS, Zaki FG, Rani I. A molecular defect in hepatic cholesterol biosynthesis in sitosterolemia with xanthomatosis. *J Clin Invest* 1990;86: pp. 923–31.

[450] Salen G, Horak I, Rothkopf M, Cohen JL, Speck J, Tint GS, Shore V, et al. Lethal atherosclerosis associated with abnormal plasma and tissue sterol composition in sitosterolemia with xanthomatosis. *J Lipid Res* 1985;26: pp. 1126–33.

[451] Acton S, Rigotti A, Landschulz KT, Xu S, Hobbs HH, Krieger M. Identification of scavenger receptor SR-BI as a high density lipoprotein receptor. *Science* 1996;271: pp. 518–20.

[452] Wang DQ, Carey MC. Susceptibility to murine cholesterol gallstone formation is not affected by partial disruption of the HDL receptor SR-BI. *Biochim Biophys Acta* 2002;1583: pp. 141–50.

[453] Kozarsky KF, Donahee MH, Rigotti A, Iqbal SN, Edelman ER, Krieger M. Overexpression of the HDL receptor SR-BI alters plasma HDL and bile cholesterol levels. *Nature* 1997;387: pp. 414–7.

[454] Eckhardt ER, Moschetta A, Renooij W, Goerdayal SS, van Berge-Henegouwen GP, van Erpecum KJ. Asymmetric distribution of phosphatidylcholine and sphingomyelin between micellar and vesicular phases. Potential implications for canalicular bile formation. *J Lipid Res* 1999;40: pp. 2022–33.

[455] van Erpecum KJ, Carey MC. Influence of bile salts on molecular interactions between sphingomyelin and cholesterol: relevance to bile formation and stability. *Biochim Biophys Acta* 1997;1345: pp. 269–82.

[456] Moschetta A, vanBerge-Henegouwen GP, Portincasa P, Renooij WL, Groen AK, van Erpecum KJ. Hydrophilic bile salts enhance differential distribution of sphingomyelin and phosphatidylcholine between micellar and vesicular phases: potential implications for their effects in vivo. *J Hepatol* 2001;34: pp. 492–9.

[457] Smit JJ, Schinkel AH, Oude Elferink RP, Groen AK, Wagenaar E, van Deemter L, Mol CA, et al. Homozygous disruption of the murine mdr2 P-glycoprotein gene leads to a complete absence of phospholipid from bile and to liver disease. *Cell* 1993;75: pp. 451–62.

[458] Oude Elferink RP, Beuers U. Targeting the ABCB4 gene to control cholesterol homeostasis. *Expert Opin Ther Targets* 2011;15: pp. 1173–82.

[459] Oude Elferink RP, Paulusma CC. Function and pathophysiological importance of ABCB4 (MDR3 P-glycoprotein). *Pflugers Arch* 2007;453: pp. 601–10.

[460] Langheim S, Yu L, von Bergmann K, Lutjohann D, Xu F, Hobbs HH, Cohen JC. ABCG5 and ABCG8 require MDR2 for secretion of cholesterol into bile. *J Lipid Res* 2005;46: pp. 1732–8.

[461] Dikkers A, Tietge UJ. Biliary cholesterol secretion: more than a simple ABC. *World J Gastroenterol* 2010;16: pp. 5936–45.

[462] Lammert F, Wang DQ, Hillebrandt S, Geier A, Fickert P, Trauner M, Matern S, et al. Spontaneous cholecysto- and hepatolithiasis in Mdr2-/- mice: a model for low phospholipid-associated cholelithiasis. *Hepatology* 2004;39: pp. 117–28.

[463] Davit-Spraul A, Gonzales E, Baussan C, Jacquemin E. The spectrum of liver diseases related to ABCB4 gene mutations: pathophysiology and clinical aspects. *Semin Liver Dis* 2010;30: pp. 134–46.

[464] Gonzales E, Davit-Spraul A, Baussan C, Buffet C, Maurice M, Jacquemin E. Liver diseases related to MDR3 (ABCB4) gene deficiency. *Front Biosci* 2009;14: pp. 4242–56.

[465] Paulusma CC, Elferink RP, Jansen PL. Progressive familial intrahepatic cholestasis type 1. *Semin Liver Dis* 2010;30: pp. 117–24.

[466] Jansen PL, Sturm E. Genetic cholestasis, causes and consequences for hepatobiliary transport. *Liver Int* 2003;23: pp. 315–22.

[467] Cohen DE, Leonard MR, Carey MC. In vitro evidence that phospholipid secretion into bile may be coordinated intracellularly by the combined actions of bile salts and the specific phosphatidylcholine transfer protein of liver. *Biochemistry* 1994;33: pp. 9975–80.

[468] Fuchs M, Carey MC, Cohen DE. Evidence for an ATP-independent long-chain phosphatidylcholine translocator in hepatocyte membranes. *Am J Physiol* 1997;273: pp. G1312–9.

[469] Erlinger S: Bile Flow. In: Arias IM, Jakoby WB, Popper H, D. S, A. SD, eds. The Liver: Biology and Pathobiology. 2 ed. New York: Raven Press, 1988; pp. 643–61.

[470] Holzbach RT. Metastability behavior of supersaturated bile. *Hepatology* 1984;4: pp. 155S–8S.

[471] Holzbach RT, Marsh M, Olszewski M, Holan K. Cholesterol solubility in bile. Evidence that supersaturated bile is frequent in healthy man. *J Clin Invest* 1973;52: pp. 1467–79.

[472] Olszewski MF, Holzbach RT, Saupe A, Brown GH. Liquid crystals in human bile. *Nature* 1973;242: pp. 336–7.

[473] Holzbach RT, Corbusier C. Liquid crystals and cholesterol nucleation during equilibration in supersaturated bile analogs. *Biochim Biophys Acta* 1978;528: pp. 436–44.

[474] Halpern Z, Dudley MA, Kibe A, Lynn MP, Breuer AC, Holzbach RT. Rapid vesicle formation and aggregation in abnormal human biles. A time-lapse video-enhanced contrast microscopy study. *Gastroenterology* 1986;90: pp. 875–85.

[475] Halpern Z, Dudley MA, Lynn MP, Nader JM, Breuer AC, Holzbach RT. Vesicle aggregation in model systems of supersaturated bile: relation to crystal nucleation and lipid composition of the vesicular phase. *J Lipid Res* 1986;27: pp. 295–306.

[476] Holzbach RT, Pak CY. Metastable supersaturation. Physicochemical studies provide new insights into formation of renal and biliary tract stones. *Am J Med* 1974;56: pp. 141–3.

Author Biographies

Dr. David Q.-H. Wang is an Associate Professor of Internal Medicine at Saint Louis University School of Medicine, St. Louis, MO. In 1992–1996, as a postdoctoral research fellow, he studied the pathophysiology of cholesterol gallstone formation, the physical-chemistry of lipids, and the molecular biochemistry of cholesterol and bile acid metabolism with Professor M.C. Carey at Harvard Medical School and Brigham and Women's Hospital, Boston, MA. At the same time, he studied the genetic mechanism of gallstone formation with Dr. B. Paigen at the Jackson Laboratory, Bar Harbor, ME. In 1996, he joined the faculty at Harvard Medical School, first as an Instructor of Medicine (1996–2000) and then an Assistant Professor of Medicine (2001–2010). Also, he worked as Gastroenterologist at Beth Israel Deaconess Medical Center and Harvard Medical School, Boston, MA (2000–2010). Dr. Wang's major research interest is focused on the pathophysiology and genetics of cholesterol gallstone disease, and the physiology and genetics of intestinal absorption of cholesterol and fatty acids, the physical-chemistry of cholesterol crystallization in bile, and the pathophysiology of cholestasis, nonalcoholic fatty liver disease, and the metabolic syndrome. His research has been supported by NIH, foundations, and industry. Dr. Wang received the Industry Scholar Research Award from the American Digestive Health Foundation and American Gastroenterological Association (1996–1999) and the New Scholar Award from the Ellison Medical Foundation (1999–2003). He has published over 80 peer-reviewed papers.

Dr. Neuschwander-Tetri is a Professor of Internal Medicine and the Director of the Division of Gastroenterology and Hepatology at Saint Louis University, St. Louis, MO. He completed his undergraduate studies at the University of Oregon and received his M.D. degree from Yale University. After internship and residency in internal medicine at the University of Wisconsin Madison, he completed his fellowship in gastroenterology and hepatology at University of California San Francisco. In 1991 he joined the faculty at Saint Louis University where he conducts clinical and basic research in nonalcoholic steatohepatitis. He is a participant in the NIDDK NASH Clinical Research Network which is conducting multicenter studies to understand the causes and identify treatments for NASH. He is actively involved in the teaching programs for the medical students, residents, and subspecialty residents, and he supervises the care of patients with advanced liver disease in inpatient and outpatient settings.

Dr. Piero Portincasa is a Professor of Internal Medicine at the University Medical School of Bari, Italy. In 1985–1987, as Research Fellow, he studied at Guy's Hospital, London, UK. In 1991 he settled as Assistant Professor at the University of Bari Medical School. In 1993–1995, as Internist from the University of Bari, he was Research Fellow and Staff Member at the Academic Hospital at the University in Utrecht, the Netherlands, where he completed his Ph.D. studies. In 2001 he was nominated Associate Professor of Internal Medicine at the University of Bari. In 2010 he was selected as Full Professor of Internal Medicine, at the national competition at the University of Bologna, Italy. He is an active member of several International Scientific Societies, and Journal Editorial Boards, Member of the Apulian Academy of Sciences (2009), President of the European Society for Clinical Investigation (ESCI: 2011–14), Honorary Visiting Professor at the University of Cluj-Napoca (Romania) (since 2009), Councilor of the Apulian Section of Italian Society Internal Medicine (SIMI) (2010–11), and Honorary Member of the Romanian Society Gastroenterology and Hepatology. He has been mentor and panelist in Ph.D. programs shared with the University of Utrecht (The Netherlands), Coimbra and Porto (Portugal), and Saint Louis (USA). He has been active in international relations, and since 2009, has been Faculty Delegate for the Erasmus/Long Life Learning Program for transeuropean mobility of teachers and students. Dr. Portincasa's major research interest is in the area of lipid metabolism and enterohepatic circulation with respect to mechanisms leading to cholelithiasis, fatty liver, and metabolic syndrome. He has been performing several translational studies focusing on transport of water and ions in the hepato-intestinal tract, and gastrointestinal motility.